Socratic Physics

Volume I
A WORKBOOK APPROACH
TO CONCEPTS OF MECHANICS

George Mathew
Northern Kentucky University

N. N. Mathew

Brooks/Cole Publishing Company

I(T)P® An International Thomson Publishing Company

Pacific Grove • Albany • Belmont • Bonn • Boston • Cincinnati • Detroit • Johannesburg • London
Madrid • Melbourne • Mexico City • New York • Paris • Singapore • Tokyo • Toronto • Washington

Senior Developmental Editor: *Keith Dodson*
Editorial Assistant: *Nancy Conti*
Marketing Team: *Margaret Parks, Steve Catalano, Deanne Brown*
Production Editor: *Laurel Jackson*
Printing and Binding: *Patterson Printing.*
Cover Design: *Vernon T. Boes*

For more information, contact:

BROOKS/COLE PUBLISHING
511 Forest Lodge Road
Pacific Grove, CA 93950
USA

International Thomson Editores
Seneca 53
Col. Polanco
11560 México, D. F., México

International Thomson Publishing Europe
Berkshire House 168-173
High Holborn
London WC1V 7AA
England

International Thomson Publishing Japan
Hirakawacho Kyowa Building, 3F 418
2-2-1 Hirakawacho
Chiyoda-ku, Tokyo 102
Japan

Thomas Nelson Australia
102 Dodds Street
South Melbourne, 3205
Victoria, Australia

International Thomson Publishing Asia
60 Albert Street
#15-01 Albert Complex
Singapore 189969

Nelson Canada
1120 Birchmount Road
Scarborough, Ontario
Canada M1K 5G4

International Thomson Publishing GmbH
Königswinterer Strasse
53227 Bonn
Germany

Printed in the United States of America

5   4   3   2   1

ISBN 0-534-36581-7

To our parents and our wives

Acknowledgments

Our special thanks go to Joseph Straly and Dan Barth for their excellent work on this book.

We would also like to thank the following people who contributed their time and effort to this book: physics editors Keith MacAdam and Alan Kerrick; copy editors Ellen Curtin and Elizabeth Ackley; cartoonist Rick Weber; illustrators Dariuz Janczewski, Daren Crigler, Miles Inada, and Janet Creekmore; designer (with Janet Creekmore) Dan Jasper; Web page designer John J. Miller; and Penny Weed, Emily Lockhart, Courtney Bailey, Robin Moody, and Brett Scharf.

We sought permission to reproduce the contributions from Angel Joseph, Pearl D'Mello, Premy Augustus, and Michael Slaughter.

Quick Welcome Quiz!

Thank you for using our book. Good Luck!

1. An object is thrown vertically upward with a non-zero velocity. If gravity is turned off at the instant the object reaches the maximum height, what happens? Assume vertical motion.

    Wrong Answer: The object proceeds to move in a straight line.

    Correct Answer: The object will be at rest. The velocity at the maximum height just before the gravity was turned off was zero. Let us call that time $t = 0$ s. Beyond $t = 0$ s, no forces are acting on the object. By Newton's Law, it should stay at rest forever!

2. Same problem as above but now the object is thrown at an angle to the horizontal.

    Wrong Answer: It will stay at rest.

    Correct Answer: The vertical velocity in the presence of gravity at the maximum height is zero. But it still has a non-zero horizontal component at the maximum height for the velocity. This makes the object go in a straight line parallel to the x axis!

3. If a freely standing object of mass m is at rest on a horizontal surface, what is the force of friction?

    Wrong Answer: $\mu \times$ normal force $= \mu \times (mg)$.

    Correct Answer: Force of friction is zero.

4. If an object is at rest on an inclined plane, what is the force of friction?

    Wrong Answer: Zero. This happened because we discussed the correct answer to Q3 before asking Q4.

    Correct Answer: Force of friction = $mg \sin \theta$, where m is the mass in kg, and θ is the angle of the inclined plane with respect to the horizontal.

Linear Motion **1**

Projectile Motion **2**

Forces & Motion **3**

Inclined Planes... **4**

Circular Motion **5**

Note: Every THIRD problem is a homework problem. You provide the answers! Solutions to the rest are provided at the back of the book. We encourage you to read OUR solutions to 2/3's of the problems. Good Luck!

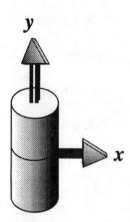

"I HAVE HEARD ABOUT NORTH, SOUTH, EAST, AND WEST BUT NOT POSITIVE EX, NEGATIVE EX, POSITIVE WHY, AND NEGATIVE WHY!"

For problems **1-5**, a car goes from station A to station B with constant speed v_1, and returns from B to A with constant speed v_2. The separation between station A and station B is d meters.

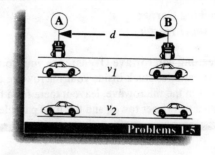

Problems 1-5

1. The total time for the trip is:

   A. $2d / (v_1 + v_2)$.
   B. $(d / v_1) + (d / v_2)$.
   C. $4d / (v_1 + v_2)$.
   D. $2d / (v_1 \times v_2)^{1/2}$.

2. The total distance in this case is:

   A. $2d$.
   B. Zero.

3. The total displacement is:

   A. $2d$ along (**+x**).
   B. $2d$ along (**−x**).
   C. Zero.

4. The average speed during the trip is:

   A. $(v_1 + v_2) / 2$.
   B. $(2 \times v_1 \times v_2) / (v_1 + v_2)$.
   C. $(v_1 \times v_2) / (v_1 + v_2)$.
   D. Zero.

5. The average velocity is:

   A. $(v_1 + v_2) / 2$.
   B. $(2 \times v_1 \times v_2) / (v_1 + v_2)$.
   C. $(v_1 \times v_2) / (v_1 + v_2)$.
   D. zero.

6. Stations A and B are d meters apart with station C exactly in the middle. A car goes from station A to station C at a constant speed of v_1. From station C, it then proceeds to station B at a constant speed of v_2. The average speed for the entire trip is:

   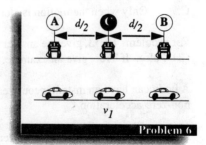

   Problem 6

   A. $(v_1 + v_2) / 2$.
   B. $(2 \times v_1 \times v_2) / (v_1 + v_2)$.
   C. $(v_1 \times v_2) / (v_1 + v_2)$.

> **Cooking Corner**
> *by Angel Joseph*
>
> **Cleaning Microwave Ovens:** Are you worried about the tough
> food particles stuck inside your microwave oven? ... Boil a cup of
> water in the microwave, leave it there for a few minutes. Wipe
> with a wet paper towel, and the food particles come off easily.

7. A person walks d meters from store A to store B. The speed is
v_1 for the first quarter distance of the trip, v_2 for the next
quarter, and v_3 for the rest. The average speed for the entire
trip is:

     A. $(4 \times v_1 \times v_2 \times v_3) / (v_2 \times v_3 + v_1 \times v_3 + 2 \times v_1 \times v_2)$.
     B. $(v_1 + v_2 + v_3) / 3$.
     C. $(v_1 + v_2 + 2 \times v_3) / 3$.
     D. $(2 \times v_1 + 2 \times v_2 + v_3) / 3$.

8. Marci walks from station A to station B in t seconds. For $t/2$
seconds, she walks at a constant speed of v_1, and for the other
half of the time, she moves at a constant speed of v_2. What is
the average speed for the entire trip?

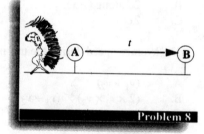
Problem 8

     A. $(v_1 + v_2) / 2$.
     B. $(2 \times v_1 \times v_2) / (v_1 + v_2)$.
     C. $(v_1 \times v_2) / (v_1 + v_2)$.
     D. $(v_1 \times v_2)^{1/2}$.
     E.   Zero.

9. The magnitude of a displacement is equal to the distance
traveled:

     A.   In all cases.
     B.   Only when the motion is along a straight line in one
            direction only.
     C.   Only when the motion is along a curved path.

For problems **10-13**, give the answer as (A) True or
(B) False. If True, give an example.

10. A body can have an instantaneous zero velocity and a
non-zero acceleration.

11. A body can have an instantaneous zero acceleration and
a non-zero velocity.

12. A body can have velocity and acceleration in different directions.

13. A body can have velocity and acceleration in the same direction.

14. A dancing couple covers one-half of the circumference of a circle of radius R. The distance traveled and the displacement of the couple are respectively:

   A. $\pi \times R$ and $\pi \times R$.
   B. $2 \times R$ and $\pi \times R$.
   C. $\pi \times R$ and zero.
   D. $\pi \times R$ and $2 \times R$.

Problem 14

15. In the absence of air resistance, a freely falling body near earth's surface, but not in a region of non-uniform gravity, experiences:

   A. A constant velocity.
   B. A constant acceleration.
   C. An increasing acceleration.
   D. A decreasing acceleration.

Problem 15

16. When an object moves in a circle with constant speed:

   A. Acceleration is zero.
   B. Velocity is a constant.
   C. Acceleration is non-zero.

   For problems **17-26**, GUESSTIMATE! Mark (A) for True and (B) for False.

17. An average speed of 75 miles/hour is roughly 33 m/s.

18. An airplane can fly at an average speed of 500 miles /hour.

19. A person walks at about 5 miles/hour when in a hurry.

20. Light takes 1 year to reach the earth from the sun.

21. If we ignore air resistance, the instantaneous speed of an object released from rest after one second during free fall near earth is roughly 10 m/s.

22. The gain in speed per second for a freely falling object is 20 m/s on the surface of the earth.

23. Sometimes winds can blow at a speed of 15 miles/hour.

24. While biking, the average speed can be 15 miles/hour.

25. Andre Agassis tennis serve can have an average speed of 125 miles/hour.

26. The average speed of the earth around the sun is roughly 20,000 m/s.

27. Instantaneous velocity _____ the average velocity.

   A.   Is always greater than.
   B.   Is always less than.
   C.   Is always equal to.
   D.   Can be greater than, less than, or equal to.

Problem 24

28. Rock Ron walks along the path A–B–C in a time of 80 minutes. The distance from A to B = the distance from B to C = 40 m. The average velocity and speed of Rock Ron are respectively:

   A.   60 m/hr and 60 m/hr.
   B.   60 m/hr along AC the "diagonal" and 60 m/hr.
   C.   56.6 m/hr along AC the "diagonal" and 60 m/hr.
   D.   42 m/hr along AC the "diagonal" and 60 m/hr.
   E.   42 m/hr along AC the "diagonal" and 42 m/hr.

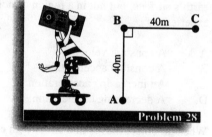

Problem 28

29. If a bus maintains constant speed, can we conclude that acceleration is zero?

   A.   Yes.
   B.   No.

30. If a bus maintains constant velocity, can we conclude that the acceleration is zero?

   A.   Yes.
   B.   No.

31. If a car maintains constant velocity of 60 mi/hr for 20 s, what is its acceleration?

   A. $1{,}200 \ mi/s^2$.
   B. $3 \ mi/s^2$.
   C. $0 \ mi/s^2$.

Circuit Breakers: *My first car was a 1972 Chevy Nova that a friend gave me in 1990. It rained on the inside, and I could always see the road when I looked down. The bumper was tied on to the chassis by a thick wire. When I went for my driver's test in Ohio, the inspector who climbed in to evaluate me looked slightly uneasy. I passed the parallel parking test, but failed the road test. The car drove better than I did, however.*

32. For a freely falling rain drop very near the surface of the earth (in the absence of air resistance), its:

   A. Speed increases.
   B. Velocity increases.
   C. Acceleration increases.
   D. Speed and velocity decrease.
   E. Speed and velocity increase at constant acceleration.

Problem 32

33. The increase in speed per second for a freely falling object vertically down (in the absence of air resistance) is about:

   A. 0 m/s^2.
   B. 0 m/s.
   C. 5 m/s^2.
   D. 10 m/s.
   E. 20 m/s.

34. An apple is in free fall straight downward. At $t = 3$ s, a speedometer fitted on the apple registers 30 m/s. What is its speed at $t = 4$ s? Ignore air resistance.

   A. 30 m/s.
   B. 35 m/s.
   C. 40 m/s.
   D. 10 m/s.

35. A yellow rose is in free fall for 3 s. If it was dropped from rest (at the origin, $x = 0$ m, $y = 0$ m), find its position at $t = 1$ s, 2 s, and 3 s. Ignore air resistance.

   A. $x = 0$ m, $y = -10$ m, -20 m, and -30 m.
   B. $x = 0$ m, $y = -5$ m, -15 m, and -25 m.
   C. $x = 0$ m, $y = -5$ m, -20 m, and -45 m.

36. A skier was seen to have velocities of 10 mi/hr, 25 mi/hr, 40 mi/hr, 55 mi/hr ... at $t = 1$ s, 2 s, 3 s, 4 s ... What is the acceleration? Is it a constant?

   A. Acceleration is zero.
   B. Acceleration is constant at 15 mi/hr.
   C. Acceleration is a constant at 15 mi/hr per second.
   D. Acceleration is not a constant.

Problem 36

37. If acceleration is a constant, we can then conclude that:

   A. Velocity is also a constant.
   B. Distances traveled for successive time intervals are the same.
   C. Change in velocity per second is not a constant.
   D. Change in velocity per second is also a constant.

38. Two cars have the same maximum speed. Which will you choose to win in a race over a short distance?

   A. The one with the maximum acceleration.
   B. The one with the minimum acceleration.
   C. It does not matter.

Problem 38

39. If two objects of different mass are released from rest in a vacuum, which object will reach the ground first?

   A. The heavier object reaches the ground first.
   B. The lighter object reaches the ground first.
   C. Both objects reach the ground at the same time.

Problem 39

40. A car accelerates from rest uniformly at 5 m/s per second. What is its speed after 5 seconds? Assume acceleration and displacement are in the same direction.

   A. 1 m/s.
   B. 2 m/s.
   C. 25 m/s.

41. A car is initially moving at a constant velocity of 4 m/s. Suddenly it accelerates uniformly at 3 m/s per second. What is its speed after 5 seconds? Assume acceleration and displacement are in the same direction.

   A. 10 m/s.
   B. 19 m/s.
   C. 25 m/s.

42. A giant truck and a small car moving at the same instantaneous velocity skid with the same deceleration and come to a stop. Which travels farther?

Problem 42

    A.   Giant truck.
    B.   Small car.
    C.   Both travel the same distance before coming to a stop.

43. An object can have a change in velocity even though it is moving at constant speed.

    A.   True.
    B.   False.

44. We are given velocities of 2 m/s and − 2 m/s along x. Which velocity is bigger in magnitude?

    A.   2 m/s.
    B.   − 2 m/s.
    C.   Both are equally big.

Figures **45–48** show displacement–time graphs of a particle. Express the state of motion as one of the following.

    A.   It has uniform deceleration.
    B.   It moves with constant acceleration.
    C.   It moves at constant velocity.
    D.   It is at rest.

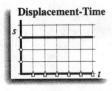

Problem 45

Problem 46

Problem 47

Problem 48

49.   The instantaneous acceleration of a body moving in a straight
       line can be found from:

       A.   The area of the velocity —time graph.
       B.   The shape of the velocity —time curve.

50.   For a book dropped from rest, which figure gives the
       velocity —time curve? Assume "down" to be negative y.

       A.   Figure A.
       B.   Figure B.
       C.   Figure C.
       D.   Figure D.

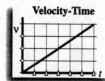

Problem 50, Figure A

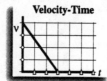

Problem 50, Figure B

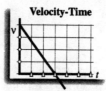

Problem 50, Figure C

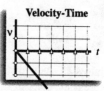

Problem 50, Figure D

For problems 51 —54 choose the correct answer from the
following.

       A.   Displacement, Δx.
       B.   Rotational speed.
       C.   Initial speed, v_0, at $t = 0$ s.
       D.   Final speed, v_f.
       E.   Acceleration, a.

51.   In the velocity —time graph (as shown in the figure), what
       quantity is represented by OA?

52.   By CB?

53.   By $\tan \theta$?

54.   By area OCBAO?

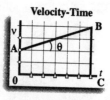

Problems 51-54

55. For a ball that is thrown vertically upward, which diagram gives the correct velocity–time curve? Ignore air resistance.

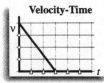

Velocity-Time

Problem 55, Figure A

Velocity-Time

Problem 55, Figure B

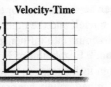

Velocity-Time

Problem 55, Figure C

Velocity-Time

Problem 55, Figure D

56. For a ball that is thrown vertically up, which figure in problem 55 best represents the *speed*-time graph? Ignore air resistance.

57. Point A is placed on the (vertical) diameter, at the maximum height. If we connect chords from point A to points B and C, and drop a bead at point A *from rest* (the bead is sliding along the chord), what can we conclude about times taken by the bead to go from point A to point B, t_{AB}, and point A to point C, t_{AC}? Ignore effects of friction.

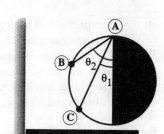

A. t_{AB} is less than t_{AC}.
B. t_{AB} is equal to t_{AC}.
C. t_{AB} is greater than t_{AC}.
D. It really depends upon the length of the chord.

58. The instantaneous velocity of a particle moving along a straight line can be found from:

A. The shape of the velocity–time graph.
B. The shape of the position–time graph.
C. The area of the velocity–time graph.
D. The area of the space–time graph.

59. A helicopter takes off vertically at a constant speed of v_{0y}, during which the pilot drops his cigarette butt. The cigarette butt will:

A. Initially go up with v_{0y}, reduce its speed, come to rest, and then increase its speed as it comes down.
B. Not go up, but will gain speed as it drops to ground.
C. Drop to the ground at constant speed.

Problem 59

Pay No Attention to That Man Behind the Curtains
by Dan Barth

Sumptuary laws (legislation aimed at controlling fashion) and customs have historically been seen as a way for the ruling class to draw a definitive line between itself and the other classes, particularly the latest group to come into influence, or to separate itself from foreign "corruption." Elizabeth I of England (reigned 1558-1603), upset with foreign influences, on several occasions restricted the length of the starched collars her courtiers favored. Decades later her Spanish counterpart, Philip IV of Spain (reigned 1621-1665) refused to receive any foreigner in court unless properly dressed in black.

60.   An airplane flies east horizontally at 400 mi/hr. If it drops a packet at an altitude of 10,000 m:

    A.   The packet will have zero speed along the eastward and downward directions as the packet is launched.

    B.   The packet will have a horizontal speed of 400 mi/hr, but it will not have any vertical speed at the start.

    C.   The packet will have a speed of 400 mi/hr in both eastward and downward directions at the start.

For problems **61-65**, an object is thrown vertically upward (from the ground) with a velocity v_{0y}.

61.   The maximum height reached, h, is:

    A. v_{0y}^2 / g.
    B. $v_{0y}^2 / (2 \times g)$.
    C. $2v_{0y}^2 / g$.
    D. $v_{0y}^2 / (4 \times g)$.

Problems 61-65

62.   The total displacement when the object returns to the ground is:

    A. $2h$ downward.
    B. $2h$ upward.
    C.   Zero.
    D. $h \times 2^{1/2}$.

63. The time taken to reach the maximum height is:

    A. $(2 \times v_{0y}) / g$.
    B. v_{0y} / g.
    C. $v_{0y} / (2 \times g)$.
    D. $v_{0y} / (4 \times g)$.

64. The time taken to reach the ground from the initial position at $t = 0$ s is:

    A. $(2 \times v_{0y}) / g$.
    B. v_{0y} / g.
    C. $v_{0y} / (2 \times g)$.
    D. $v_{0y} / (4 \times g)$.

65. If gravity is turned off at the instant the object reaches the maximum height, what happens? Assume vertical motion.

    A.   The object proceeds to move in a straight line.
    B.   The object will be at rest forever.
    C.   The object will fall back to the ground.

66. If a stone dropped from rest covers a distance s, during the last second of its fall, what is the time taken by the stone to reach the ground?

    A. $(s / g) + (1 / 2)$.
    B. $(s / g) - (1 / 2)$.
    C. s / g.
    D. $s / (2g)$.

67. Which of the following statements is *not* true?

    A.   Velocity and acceleration can have the same sign.
    B.   Velocity and acceleration can have opposite signs.
    C.   Displacement and velocity can have the same sign.
    D.   Acceleration and velocity are always in the same direction.

68. A train of length "*l*" crosses a bridge of length "*L*," at uniform speed *v*. Crossing, from the cowcatchor at the outset till the end of the caboose gets across, takes how much time?

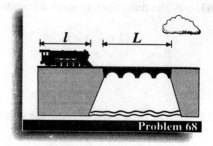
Problem 68

- A. $(L + l + l)/v$.
- B. $(L + l)/v$.
- C. $(L - l - l)/v$.
- D. $(L - l)/v$.

69. A car traveling at a speed of 100 km/hr slows down to 75 km/hr in 10 s. Choose one correct statement:

- A.   Acceleration of the car is 2.5 km/hr per second.
- B.   Deceleration of the car is 2.5 km/hr per second.

Circuit Breakers: *While at Kent State University, I decided to take a long walk, through the back roads. After a few minutes, a black dog joined me. I petted it and it stayed beside me for more than fifteen miles (wow!) until I stopped at a Burger King for a hot chocolate.*

70. The velocity-time graph for linear motion of a particle is as shown. The distance traveled during 14 seconds is:

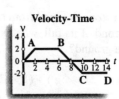
Problems 70-71

- A.   4 m.
- B.   8 m.
- C.   6 m.
- D.   2 m.
- E.   – 6 m.

71. The particle whose velocity-time curve is given in the figure for problem **70**, has zero acceleration along:

- A.   0A.
- B.   AB.
- C.   BC.

For problems **72–73**, a stone is thrown vertically upwards with a velocity, v_{0y}, from the top of a cliff of height, *h*. It reaches the bottom of the cliff in a time, *t* seconds.

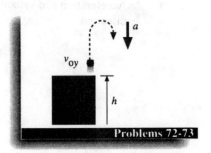
Problems 72-73

72. The total displacement is (assume **+y** as upward):

- A.   0 m.
- B.   – *h*.
- C.   + *h*.
- D. $[h - (1/2) \times g \times t^2]$.

73. The initial speed, v_{0y}, should be:

A. $[(1/2) \times g \times t^2 - h] / t$.
B. $(2 \times g \times h)^{1/2}$.
C. $[(1/2) \times g \times t]$.
D. $[(1/2) \times g \times t^2 + h] / t$.
E. $[h - (1/2) \times g \times t^2] / t$.

74. James Bond wants to jump onto a boat moving with uniform speed v_{0x}, from the top of a bridge of height, h. If he steps off the bridge, what is the time taken to land on the boat? Assume b (not shown) is the horizontal separation between him and the boat. Ignore air resistance.

Problem 74

A. $(2 \times h)^{1/2} / g$.
B. $h^{1/2} / g$.
C. h / g.
D. $(gh)^{1/2}$.

75. If James Bond is to land on the boat in problem **74**, what should be the speed of the boat?

A. $b / (2 \times h)^{1/2} / g$.
B. $b / (h^{1/2} / g)$.
C. $b / (h / g)$.
D. $b / (gh)^{1/2}$.

For problems **76–78**, two frictionless chords, AB and AC, are as shown. A bead slides along AB and AC. (Up is toward the top of the figure).

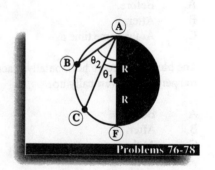

Problems 76–78

76. What is the component of acceleration due to gravity along AB, and along AC?

A. $g \times \sin \theta_1$.
B. $g \times \cos \theta_1$.
C. $g \times \tan \theta_1$.

77. If a bead is made to slide on these frictionless chords, it takes t_1 seconds for it to reach point C from A. What is the time, t_1?

A. $[2 \times (R / g)]^{1/2}$.
B. $[4 \times (R / g)]^{1/2}$.
C. R / g.
D. $2R / g$.

78.   What can you say about the times taken by the bead along AC, and along AB?

   A.   It takes more time to go from A to C than from A to B.

   B.   It takes less time to go from A to C than from A to B.

   C.   It takes the same time to go from A to C as from A to B.

For problems **79-81**, a red pen is thrown vertically up, a blue pen is thrown horizontally, and a black pen is just dropped. The black pen accelerates at 10 m/s/s downward. Ignore air resistance.

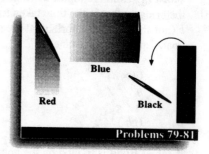

Problems 79-81

79.   The magnitude of the acceleration for the blue pen along the vertical is:

   A.   5 m/s /s.
   B.   10 m/s.
   C.   5 m/s.
   D.   10 m/s /s.

80.   The blue pen thrown horizontally reaches the ground _____ the black pen.

   A.   Before.
   B.   After.
   C.   At the same time as.

81.   The blue pen thrown horizontally reaches the ground __*C*__ the red pen thrown vertically up.

   A.   Before.
   B.   After.
   C.   At the same time as.

82.   During each second, the distance traveled by a freely falling object (assume downward motion) _____.

   A.   Keeps on increasing.
   B.   Keeps on decreasing.
   C.   Stays the same.

83. A car can accelerate at 3 m/s per second. If the acceleration is constant, what is the distance traveled in 4 s, assuming the car starts from rest?

    A.   6 m.
    B.   12 m.
    C.   18 m.
    D.   24 m.

84. A car is traveling at a constant velocity of 10 m/s. It then begins accelerating (along the direction of motion) at a constant 3 m/s per second. What is the velocity of the car after 2 s?

    A.   4 m/s.
    B.   8 m/s.
    C.   10 m/s.
    D.   16 m/s.

85. A car is traveling at a constant velocity of 10 m/s. It then begins decelerating (against the direction of motion) at a constant 3 m/s per second. What is the velocity of the car after 2 s?

    A.   4 m/s.
    B.   8 m/s.
    C.   10 m/s.
    D.   12 m/s.

86. After seeing a police car, a driver decelerates his car at 5 m/s/s to a speed of 4 m/s in a time of 2 s. What was his original speed?

    A.   14 m/s.
    B.   12 m/s.
    C.   8 m/s.
    D.   6 m/s.

87. A chalk is tossed straight up at a speed of 20 m/s. How long does it take by the chalk to reach its maximum height?

    A.   1 s.
    B.   2 s.
    C.   3 s.
    D.   4 s.

88. In problem **87**, if the chalk is caught on the way back, what is the total time elapsed (from start to finish)?

    A.   1 s.
    B.   2 s.
    C.   3 s.
    D.   4 s.

Problem 89

89. What is the launching speed of an object shot straight up if it took only one second to reach the maximum height?

    A.   10 m/s.
    B.   20 m/s.
    C.   30 m/s.
    D.   40 m/s.

90. Some bullets can be fired vertically up at a velocity of 400 m/s. After one second, what is the distance traveled in the <u>absence</u> of gravity?

    A.   400 m.
    B.   (400 + 5) m.
    C.   (400 − 5) m.
    D.   5 m.

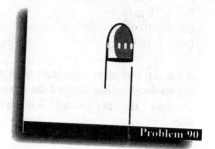

Problem 90

91. A drinking glass is dropped to the ground from a height of 1.8 m. What is its speed just before it hits the ground?

    A.   4 m/s.
    B.   4 m/s/s.
    C.   6 m/s.
    D.   6 m/s/s.

y

x

92. The vertical height for a hang time (start your stop-watch when the person is air-bound and stop it when the person touches the ground) of 2 s is:

    A.   10 m.
    B.   20 m.
    C.   5 m.
    D.   15 m.

For problems **93-100**, mark (A) for True and (B) for False.

93. An object can have an acceleration while going at constant speed.

94.   An object can have an acceleration while going at constant velocity.

95.   An object can have an acceleration in the opposite direction to velocity.

96.   An object can have an acceleration in the opposite direction to displacement.

97.   When a ball is thrown vertically up, the time required for rising is different from falling, if we ignore air resistance.

98.   A red ball thrown vertically up at 10 m/s will impact the ground at a higher speed than a blue ball thrown vertically down at 10 m/s from the top of a building of height, h.

99.   Acceleration of normal cars does not usually exceed 10 m/s/s.

100. A ball bounces off a wall at the same speed at which it approaches the wall. The change in velocity is zero.

101. Can we use equations of kinematics for all velocity–time graphs?

      A.   Yes for all cases.
      B.   Yes for segments of constant acceleration.

102. For the displacement–time graph shown, velocity is maximum at:

      A. t_1.
      B. t_2.
      C. t_3.

103. For the displacement–time graph shown, find the average velocity for the interval $t = 2$ s to $t = 5$ s.

      A.   2.7 m/s.
      B.   – 2 m/s.
      C.   4 m/s.
      D.   – 4 m/s.

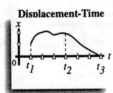

Displacement-Time

Problem 102

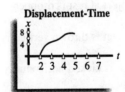

Displacement-Time

Problem 103

For figures **104-106**, choose one of the following answers about acceleration, *a*.

A.   *a* is positive.
B.   *a* is negative.
C.   *a* is zero.

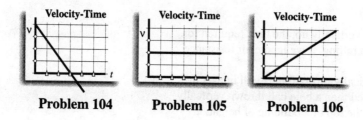

Problem 104   **Problem 105**   **Problem 106**

For problems **107 - 112**, choose one of the following velocity-time curves as your answer. Define "up" as your direction of positive displacement.

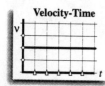 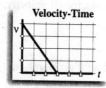 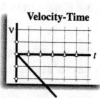

Problems 107-112, Figure A   Problems 107-112, Figure B   Problems 107-112, Figure C   Problems 107-112, Figure D

107.   An object is thrown vertically up.

108.   An object at rest is dropped.

109.   An object is thrown vertically down.

110.   A puck is kicked horizontally on a smooth frictionless surface.

111.   A car comes to rest under constant deceleration.

112.   A helicopter rises vertically at constant speed.

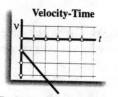

Problems 107-112, Figure E

113. A bullet is fired vertically up at a velocity of 400 m/s. After one second, what is the distance traveled in the <u>presence</u> of gravity?

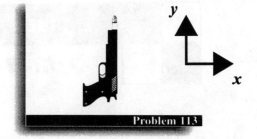

Problem 113

A. 400 m.
B. (400 + 5) m.
C. (400 − 5) m.
D. 5 m.

TRUE? **CONCEPT CHECK**

m u s t K n O w

114. Displacement is a vector.

115. Displacement can be negative with respect to a fixed axis.

116. Displacement has a magnitude and a direction.

117. For problem **6**, the time to go from A to C is $(d/2)/v_1 = d/(2v_1)$.

118. For problem **6**, the time to go from C to B is $(d/2)/v_2 = d/(2v_2)$.

119. The total time it takes to go from A to B is $d/(2v_1) + d/(2v_2)$.

120. The total distance from A to B is d.

121. Average speed = distance / total time.

122. If velocity and acceleration have the same sign, then the object speeds up (covers more distance per unit time) in succeeding time intervals.

123. If velocity and acceleration have opposite signs, then the object slows down.

124. The circumference in m of a circle is $2\pi R$, where R is the radius in m.

125. The kinematic equation $v_f = v_0 + at$ is valid only when acceleration is a constant.

126. Constant speed does not always mean constant velocity.

127. The increase in speed per second for an object freely falling vertically down is about 10 m/s.

128. An object registers 10 m/s, 20 m/s, 30 m/s, ... at intervals of 1 s while moving in a straight line. Its acceleration is 10 m/s/s.

129. An object A is always seen to cover 5 m in one second. Its average speed is then 5 m/s.

130. In a certain direction, object A covers 5 m every second, and object B covers 7 m every second. Both objects then have zero acceleration.

131. The distances covered by two objects A and B for a time interval of 4 s is as shown. Between 2 and 3 s, the average speed is the same for both objects as they cover equal distances during that time interval.

Time t in s	0-1	1-2	2-3	3-4
A	5 m	5 m	5 m	5 m
B	4 m	3 m	5 m	2 m

132. If two objects have the same acceleration and the same initial speed, they cover the same amount of distance per unit time (at a given time).

133. At maximum height (problem **61**), the vertical velocity is 0 m/s.

134. If you throw a pen along the horizontal from a height, its vertical acceleration as it falls is non-zero.

135. If you throw a pen along the horizontal from a height, its horizontal acceleration is zero. (Ignore any kind of resistance).

136. In chapter **3**, problem **85**, the masses m_1 and m_2 have the same acceleration.

137. In chapter **5**, problem **31**, the average speed of the electron for one revolution is
$v = $ circumference / time.

138. In chapter **5**, problem **42**, the total distance traveled in one second by the electron is
$6.6 \times 10^{15} \times (2\pi R)$ where R is the radius in m.

139. My daughter Samantha takes an average of 20 minutes to drink her milk from the bottle. Then 10 minutes after I give her the bottle, I can expect her to have finished half the bottle (assuming she drinks at a constant rate).

140. The automatic garage door closes fully in one minute. If you are inside and want to get out through the garage door, you will have (roughly):

time = distance to the garage door / average walking (or running) speed

to get out.

141. At a super-market, the doors that open as you enter usually move at constant velocity.

142. The next time you watch a live TV interview, note how long it takes for a person in China (far away place) to respond to a joke of a newscaster here at home. It is because the person in China gets the joke after a little time (after you heard about it by direct beaming from a shorter distance). That time being:

time ≈ distance / speed of light.

143. If the speed of *sound* in air is a constant, then we can estimate how far away the thunder took place by using the relation

distance = average speed × time.

144. We hear thunder after we see lightning even though they occur at the same time. This is because speed of light is much faster than the speed of sound in air.

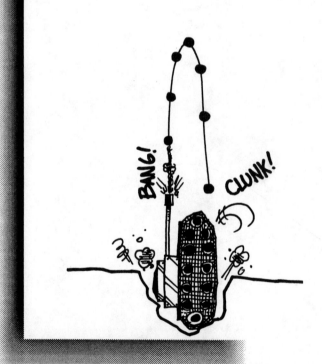

For problems **1–32**, a ball is thrown from a point O (for origin) with a velocity v_0, at an angle α with respect to the horizontal. For problems **1–6**, choose one of the answers A, B, C, D, or E. Assume there is no air resistance.

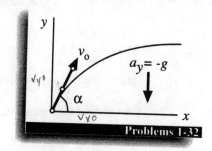

Problems 1-32

A. $v_0 \times \cos \alpha$.
B. $- g$.
C. $v_0 \times \sin \alpha$.
D. $(v_0 \times \sin \alpha) - (g \times t)$.
E. Zero.

1. What is the horizontal component of the velocity, v_{0x}, at the start?

2. What is the vertical component of the velocity, v_{0y}, at the start?

3. What is the horizontal component of the velocity, v_{fx}, after a time t?

4. What is the vertical component of the velocity, v_{fy}, after a time t?

5. What is the acceleration along the horizontal, a_x?

6. What is the acceleration along the vertical, a_y?

7. What is the magnitude of the velocity after a time t?

    A. $(v_{fx}^2 + v_{fy}^2)^{1/2}$.
    B. $v_{fx}^2 + v_{fy}^2$.
    C. $v_{fx} + v_{fy}$.
    D. v_0.

8. What is the "heading" (the direction of motion) of the velocity vector after a time t? Assume θ is the angle that the velocity vector makes with the horizontal.

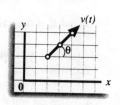

Problem 8

    A. $\tan \theta = v_{0y} / v_{0x}$.
    B. $\tan \theta = \tan \alpha$.
    C. $\tan \theta = [v_0 \times \sin \alpha - (g \times t)] / (v_0 \times \cos \alpha)$
        $= v_{fy} / v_{fx}$.

9.   What is the horizontal displacement, Δx, after a time, t?

   A. $(v_0 \times \cos \alpha) \times t$.
   B. $(v_0 \times \sin \alpha) \times t$.
   C. $-(1 / 2) \times g \times t^2$.
   D. $(v_0 \times \sin \alpha) \times t - (1 / 2) \times g \times t^2$.

10.   What is the vertical displacement, Δy, after a time, t?

   A. $(v_0 \times \cos \alpha) \times t$.
   B. $(v_0 \times \sin \alpha) \times t$.
   C. $-(1 / 2) \times g \times t^2$.
   D. $(v_0 \times \sin \alpha) \times t - (1 / 2) \times g \times t^2$.

$V_0 \cos \theta \times t$

$V \times t - \frac{1}{2} g t^2$

11.   When is the displacement, Δy, negative (i.e., the object has
        fallen below the starting point)?

   A.   If $(v_0 \times \sin \alpha) \times t$ is greater than $(1 / 2) \times g \times t^2$.
   B.   If $(v_0 \times \sin \alpha) \times t$ is less than $(1 / 2) \times g \times t^2$.
   C.   If $(v_0 \times \sin \alpha) \times t$ is equal to $(1 / 2) \times g \times t^2$.

$\frac{1}{2} g t^2 = (v_0 \sin \alpha) t$

$\frac{\frac{1}{2} g}{g}$

12.   The displacement, Δy, is zero (i.e., the object is back at the
        same level along y as the starting position, when
        $(v_0 \times \sin \alpha) = (1 / 2) \times g \times t$.

   A.   True.
   B.   False.

13.   The horizontal distance traveled when the object has returned
        to the $y = 0$ meters level is given by:

   A. $\Delta x = (v_0 \times \cos \alpha) \times [(2 \times v_0 \times \sin \alpha) / g\,]$.
   B. $\Delta x = (v_0^2 \times \cos^2 \alpha) / (2 \times g\,)$.
   C. $\Delta x = 0$ meters.

14.   In general, what is the value of the horizontal range, R?

   A. Δx, when $y = 0$ m.
   B. Δx, when $y = x / 2$ m.
   C. Δx, when $\Delta y = 0$ m.
   D. y, when $\Delta x = y / 2$ m.

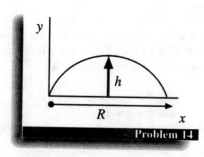
Problem 14

15.   At the maximum height, the velocity along:

   A. x but not y is zero.
   B. y but not x is zero.
   C.   both x and y is zero.

> **Chicken Soup with Sweet Corn**
> *by Pearl D'Mello*
>
> 4 pieces of boned chicken breast, 1 green pepper, 1 onion, 1 tomato, 2 cans of cream style corn, 2 egg whites, 2–3 tablespoons cornstarch, salt and black pepper to taste.
>
> In a large pot, boil the chicken with all the vegetables (coarsely chopped) in 8 cups of water for about half an hour to make a broth. Strain the broth and discard the vegetables. Shred the chicken as you would for a chicken salad and set aside. Bring the clear broth to a boil and add salt and pepper to taste. Empty the two cans of corn in the broth and reheat to boiling. In a small bowl, stir cornstarch in half a cup of water, and pour it into the broth, stirring continuously till the broth has a thick consistency. (If it becomes too thick, add a little hot water.) Allow it to simmer for a while. In the meantime, beat the egg-whites lightly with two tablespoons of water, then stir the mixture into the broth. Finally, add the shredded chicken into the soup, and serve hot with soy sauce and tabasco (for spicy flavor)

16. At the maximum height, the acceleration along:

    A. x but not y is zero.
    B. y but not x is zero.
    C.   both x and y is zero.

17. The maximum height reached, h, is:

    A. $(v_0^2 \times cos^2 \alpha) / g$.
    B. $(v_0^2 \times cos^2 \alpha) / (2 \times g)$.
    C. $(v_0^2 \times sin^2 \alpha) / g$.
    D. $(v_0^2 \times sin^2 \alpha) / (2 \times g)$.

18. What is the displacement along x when the ball reaches the maximum height?

    A. $\Delta x = (v_0 \times cos \alpha) \times [(2 \times v_0 \times sin \alpha) / g]$.
    B. $\Delta x = (v_0^2 \times cos^2 \alpha) / (2 \times g)$.
    C. $\Delta x = 0$ meters.
    D. $\Delta x = (v_0 \times cos \alpha) \times [(v_0 \times sin \alpha) / g]$.

19.   If R is the range and h is the height, then y is equal to h when x
            is equal to:

    A. R.
    B. $2 \times R$.
    C.   0.
    D. $R / 2$.

20.   What is the time required for the object to reach maximum
            height?

    A. $(v_0 \times sin\ \alpha) / g$.
    B. $(2 \times v_0 \times sin\ \alpha) / g$.
    C. $(v_0 \times cos\ \alpha) / g$.
    D. $(2 \times v_0 \times cos\ \alpha) / g$.

21.   What is the time taken by the ball to reach the ground from the
            starting point?

    A. $(v_0 \times sin\ \alpha) / g$.
    B. $(2 \times v_0 \times sin\ \alpha) / g$.
    C. $(v_0 \times cos\ \alpha) / g$.
    D. $(2 \times v_0 \times cos\ \alpha) / g$.

22.   What are the vertical and horizontal displacements when the
            ball touches the ground?

    A.   Zero and $(v_0^2 \times sin\ 2\alpha) / g$.
    B. $(v_0^2 \times sin\ 2\alpha) / g$ and zero.
    C.   Zero and $(v_0^2 \times sin\ 2\alpha) / (2 \times g)$.
    D. $(v_0^2 \times sin\ 2\alpha) / (2 \times g)$ and zero.

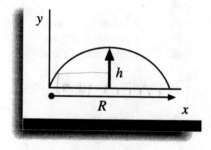

23.   What value of the projection angle, α, will achieve the
            maximum range?

    A.   15 or 75 degrees.
    B.   30 or 60 degrees.
    C.   45 degrees.
    D.   25 degrees.

24.   The horizontal displacement when the vertical height is
            $h / 2$ is:

    A.   Less than $R / 4$.
    B.   Equal to $R / 4$.
    C.   Greater than $R / 4$.

25. What is the path of the ball?

    A.   A straight line.
    B.   A hyperbola.
    C.   A parabola.
    D.   An ellipse.

26. The horizontal and vertical components of the velocity at the instant when the ball strikes the ground are:

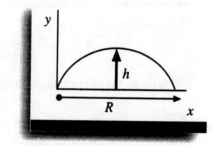

    A. $(v_0 \times cos\ \alpha)$ and $(v_0 \times sin\ \alpha)$.
    B.   Zero.
    C. $(v_0 \times cos\ \alpha)$ and $(-\ v_0 \times sin\ \alpha)$.

For problems **27–32**, choose an answer from Figures **A** through **E**. Refer to the figure here showing the horizontal range, *R*. The ball is thrown as stated before in problem **1**.

27. Which figure shows the displacement–time graph along *x*?

Problem 27

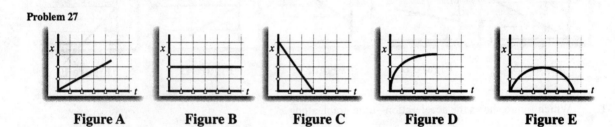

    Figure A   **Figure B**   **Figure C**   **Figure D**   **Figure E**

28. Which figure shows the displacement–time graph along *y*?

Problem 28

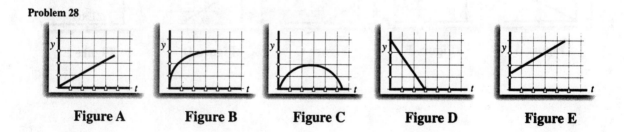

    Figure A   **Figure B**   **Figure C**   **Figure D**   **Figure E**

29. Which figure shows the velocity—time graph along *x*?

Problem 29

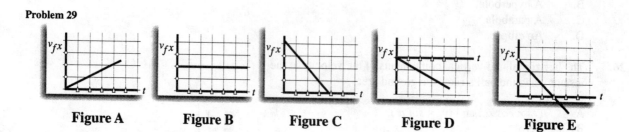

Figure A   **Figure B**   **Figure C**   **Figure D**   **Figure E**

30. Which figure shows the velocity—time graph along *y*?

Problem 30

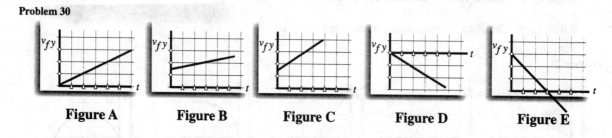

Figure A   **Figure B**   **Figure C**   **Figure D**   **Figure E**

31. Which figure shows the acceleration—time graph along *x*?

Problem 31

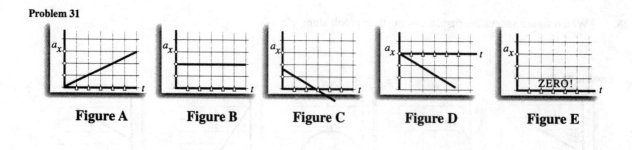

Figure A   **Figure B**   **Figure C**   **Figure D**   **Figure E**

32. Which figure shows the magnitude of the acceleration—time graph along *y*?

Problem 32

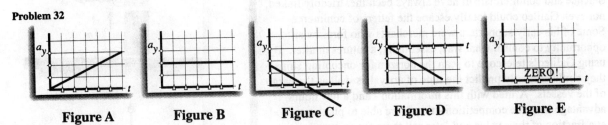

| Figure A | Figure B | Figure C | Figure D | Figure E |

33. A passenger drops an object from the window of a train moving at constant velocity on level rails. The path of the object as seen by the passenger and by an observer on the ground are:

  A. A straight line and a parabola, respectively.
  B. A parabola and a straight line, respectively.
  C. Both parabolas.
  D. Both straight lines.

34. A projectile is shot with a given initial speed. At what angle above the horizontal should it be projected to reach maximum horizontal range? Maximum height?

  A. 45 degrees, 45 degrees.
  B. 90 degrees, 90 degrees.
  C. 45 degrees, 90 degrees.

y

x

35. In order to hit a target at a height *h* meters above the ground, a rifle fired from a height of *h* meters should be aimed (assuming *h* is positive and that we have a level ground):

  A. Horizontally.
  B. At an angle above the horizontal.
  C. Vertically.
  D. At an angle below the horizontal.

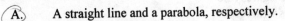

Pay No Attention To That Man Behind The Curtain
by Dan Barth

Science and commercialism have always been inextricably linked, not even Galileo could totally escape the fetters of commerce. Some of his early supporters were merchants who foresaw an opportunity to control what was, essentially, a futures market. By using Galileo's telescope to scan the horizon for oncoming ships, they could correctly predict the time of arrival as well as the cargo of the vessels. Armed with this information - and a few hours advantage over the competition - they were able to purchase goods at a fraction of their value and later sell them for an exaggerated profit.

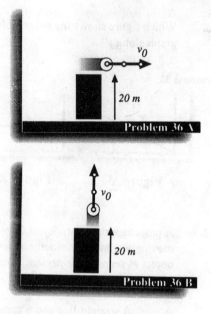

Problem 36 A

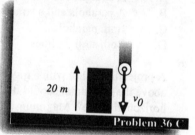

Problem 36 B

For problems **36—46**, five balls are released from a height of 20 m above the ground, each having a velocity with magnitude 10 m/s, in the directions shown in the figures (the projection angle for figures D and E is 30 degrees with respect to the horizontal). Answer the questions using the figures.

[Assume: v_{0y} = initial velocity along y; v_{0x} = initial velocity along x; h = vertical height in meters above the launching point; the heading of the velocity vector (direction of motion), $tan \phi = v_{fy} / v_{fx}$]

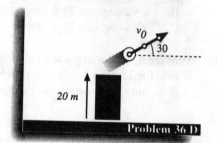

Problem 36 C

36. Which balls have the same horizontal (non—zero) speed, v_{0x}?

    A.   Figures A & B.
    B.   Figures B & C.
    C.   Figures C & D.
    D.   Figures D & E.

37. Which ball has the largest horizontal speed?

    A.   Figure A.
    B.   Figure B.
    C.   Figure C.
    D.   Figure D.
    E.   Figure E.

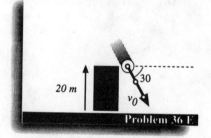

Problem 36 D

38. At the start, which balls have the largest vertical speed, v_{0y}?

    A.   Figures A & B.
    B.   Figures B & C.
    C.   Figures C & D.
    D.   Figures D & E.

Problem 36 E

39. Which ball takes the longest time to reach the ground?

   A. Figure A.
   (B.) Figure B.
   C. Figure C.
   D. Figure D.
   E. Figure E.

40. Which balls reach the ground at the same time?

   A. Figures A & B.
   B. Figures B & C.
   C. Figures C & D.
   D. Figures D & E.
   E. None of the balls reach at the same time.

41. Which ball takes the shortest time to reach the ground?

   A. Figure A.
   B. Figure B.
   C. Figure C.
   D. Figure D.
   E. Figure E.

42. Which balls hit the ground with the larger vertical velocity (in magnitude)?

   A. Figures A & B.
   B. Figures B & C.
   (C.) Figures C & D.
   D. Figures D & E.

43. Which ball hits the ground with the smaller vertical velocity (in magnitude)?

   A. Figure A.
   B. Figure B.
   C. Figure C.
   D. Figure D.
   E. Figure E.

44. Which ball travels the farthest along x?

   A. Figure A.
   B. Figure B.
   C. Figure C.
   D. Figure D.
   E. Figure E.

You Can Do It!

45. Which ball reaches the greatest height?

    A.   Figure A.
    B.   Figure B.
    C.   Figure C.
    D.   Figure D.
    E.   Figure E.

46. Which balls have parallel velocities as they strike the ground?

    A.   Figures A & B.
    B.   Figures B & C.
    C.   Figures C & D.
    D.   Figures D & E.
    E.   Figures B & C, and Figures D & E.

47. A stone is dropped from rest and takes a time of t_1 seconds to hit the ground. Another stone projected horizontally at any speed will take how long to reach the ground?

    A.   More than t_1 seconds.
    B. t_1 seconds.
    C.   Less than t_1 seconds.

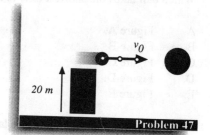

48. A ball is projected at an angle of 30 degrees with respect to the horizontal, at a certain speed, such that it strikes the ground a distance R meters from the launching point. At what other angle can we throw the ball with the same speed to reach the ground the same distance R meters?

    A.   15 degrees.
    B.   45 degrees.
    C.   60 degrees.
    D.   90 degrees.

49. R_{max} is the maximum horizontal distance a basketball can be thrown. Throwing with the same initial speed, what is its maximum vertical height? [Before it touches the ground, $\alpha = 45$. Assume same height for the start and finish.]

    A. R_{max}.
    B. $R_{max} / 2$.
    C. $2 \times R_{max}$.
    D. $R_{max} / 4$.

50. For a given velocity of projection and projection angle α, the ratio of maximum height to horizontal range is:

    A. $4 \times \tan \alpha$.
    B. $(\tan \alpha) / 4$.
    C. $\tan \alpha$.
    D. $2 \times \tan \alpha$.
    E. $(\tan \alpha) / 2$.

51. Bullets are fired from an AK—47 rifle with the same speed v_0 at all angles. Neglecting air resistance, all the bullets will fall within a circle of area:

    A. $(\pi \times v_0^4) / g^2$.
    B. $(\pi \times v_0^4) / (2 \times g^2)$.
    C. $(2 \times \pi \times v_0^4) / g^2$.
    D. $(\pi \times v_0^4) / (4 \times g^2)$.

52. What is the magnitude of acceleration of a ball sliding along the incline if the inclined plane is at an angle α to the horizontal? Assume no friction.

    A. g.
    B. $g \times \cos \alpha$.
    C. $g \times \sin \alpha$.
    D. $g \times \tan \alpha$.

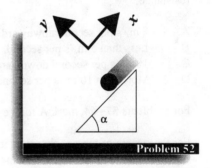

Problem 52

For problems **53—55**, assume no friction and set $\alpha = 30^0$.

53. An object starts at a velocity of 2 m/s down an inclined plane. What was its speed at the bottom of the incline, if it took 2 s to reach the bottom point?

    A. 2 m/s.
    B. 19.3 m/s.
    C. 12 m/s.
    D. 8 m/s.

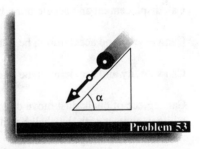

Problem 53

54. An object comes to the base of an inclined plane at a speed of 20 m/s, and starts up the incline. What is its speed after 2 s, if it climbs up the inclined plane? Assume no bounce, smooth transition.

    A. 40 m/s.
    B. 30 m/s.
    C. 10 m/s.
    D. 16 m/s.

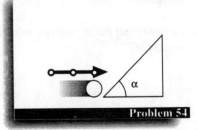

Problem 54

55.   If an object comes to the base of the inclined plane (shown in problem **54**) with a speed of 30 m/s, what distance will it travel up the inclined plane before it comes to a stop?

   A.   60 m.
   B.   70 m.
   C.   80 m.
   D.   90 m.
   E.   110 m.

56.   A car rolls down a hill with a 5—degree slope (= α). Its acceleration is 5 m/s per second, as shown in the figure, from A to B. At point B, the car falls off a cliff that is 25 m high. What is the acceleration beyond point B, in the absence of air resistance?

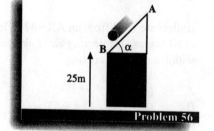

Problem 56

   A.   5 m/s per second downward.
   B.   Less than 5 m/s per second.
   C.   10 m/s per second downward.
   D.   More than 10 m/s per second.

   For problems **57—64**, mark A for Yes and B for No.

57.   For the system of masses shown, is it possible for them both to have the same acceleration? Assume no friction.

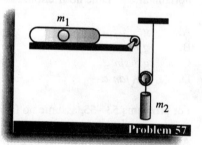

Problem 57

58.   Can displacement and acceleration be in the same direction?

59.   Can displacement and acceleration be in opposite directions?

60.   Can velocity and acceleration be in the same direction?

61.   Can velocity and acceleration be in opposite directions?

62.   Can a piece of chalk dust move downward at constant speed while in contact with the chalkboard?

63.   Can an object fall at constant velocity in the presence of air resistance?

64.   Can an object fall at constant velocity in the absence of air resistance?

65. For the system of masses shown, what is the acceleration of m_2 if the acceleration of m_1 is 5 m/s^2? Assume m_1 and m_2 are free to move, and that m_2 is attached to a ceiling.

   A.   5 m/s^2.
   B.   2.5 m/s^2.
   C.   7.5 m/s^2.
   D.   10 m/s^2.

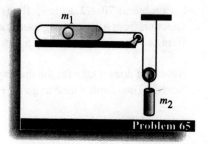
Problem 65

66. If a pitcher throws a ball 10 m before it hits ground, throwing horizontally from a height of 2 m, what is the player's pitching speed?

   A.   15.8 m/s.
   B.   5.8 m/s.
   C.   9 m/s.
   D.   2 m/s.

67. A bullet is fired horizontally with a velocity of 400 m/s, from a height of 20 m above the ground. At what distance from the gun barrel will the bullet strike the ground?

   A.   200 m.
   B.   400 m.
   C.   600 m.
   D.   800 m.

68. A projectile is fired from a chosen origin. In the presence of gravity, what is the velocity along y at two points A and B of the same height from a horizontal ground?

   A.   The same in magnitude, but opposite in direction.
   B.   The same.

69. A projectile is fired from the origin. In the presence of gravity, if speeds along the vertical at points A and B measured 20 m/s, then: [Assume up is +y.]

   A.   The object has a velocity component of -20 m/s at point A along the vertical.
   B.   The object has a velocity component of +20 m/s at point A along the vertical.
   C.   We just can't tell whether the object has a positive velocity or not.

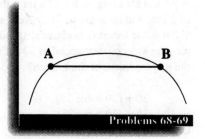
Problems 68-69

For problems **70—72**, a clown flings a dummy in the vertical direction from a car moving east at a constant velocity of 10 m/s. The dummy rises by 5 m to its highest point.

70. How long does it take for the dummy to return into the clown's hands? [Total time = time to go up + time to come down.]

    A.   1 s.
    B.   2 s.
    C.   3 s.
    D.   4 s.

71. Which one covers more horizontal distance?

    A.   Clown.
    B.   Dummy.
    C.   Neither; both travel the same.

72. The distance traveled by the clown between throw and catch is:

    A.   10 m.
    B.   20 m.
    C.   30 m.
    D.   40 m.
    E.   50 m.

73. A person standing on top of a train throws a tomato at a speed of 20 m/s at an angle of 75 degrees above the horizontal. What is the tomato's horizontal speed after 1 s, if the train is moving at a constant speed of 30 m/s in the same direction as the displacement of the tomato?

    A. $30 + (20 \times cos\ 75)$.
    B. $30 + (20 \times sin\ 75)$.
    C. $30 - (20 \times cos\ 75)$.
    D. $30 - (20 \times sin\ 75)$.

74. For problems **74-76**, meet Spanky, the stunt driver. He intends to clear a 12-m high stack of cars after being airborne for 3 s. If he is to take off from a 3-m tall ramp, what should be his vertical take-off speed?

   A. 100 m/s.
   B. 50 m/s.
   C. 36 m/s.
   D. 18 m/s.
   E. 11 m/s.

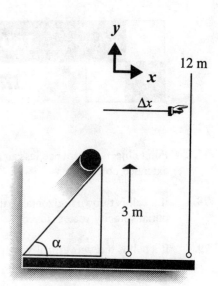

75. Yes! Spanky cleared the 12-m high stack of cars! How far horizontally from the ramp did he place the stack of cars, if his horizontal launch speed was 12 m/s?

   A. 100 m.
   B. 50 m.
   C. 36 m.
   D. 18 m.
   E. 11 m.

76. If you assume the car to be a point object, what was the initial projection angle $\theta = \alpha$ for the car?

   A. 16.3°.
   B. 26.3°.
   C. 36.3°.
   D. 46.3°.
   E. 56.3°.

77. Point objects experience less air resistance compared to extended objects.

78. If a ball is thrown horizontally, the vertical component of the initial velocity vector is zero.

79. If a ball is thrown vertically up, the horizontal component of the initial velocity vector is zero.

80. For problem **36**, Figures **A-E**, the vertical displacement between the initial launching point and the final landing point just before the projectile hits the ground, is the same in all figures.

81. For problem **36**, Figures **A-E**, the vertical displacement is $-h$ between the initial launching point and the final landing point just before the projectile hits the ground.

82. For problem **47**, the vertical displacement is negative.

83. For problem **36**, Figure **B**, the horizontal displacement is zero between the initial launching point and the final landing point (assume it starts at 0 m which is really not necessary as displacement is relative).

84. For problem **36**, Figure **B**, the vertical displacement is positive if the initial point of consideration is the launching point and the final point is the maximum height.

85. For problem **36**, Figure **B**, the vertical displacement is zero if the initial point on the way up is at the same height as the final point on the way down.

86. For problem **36**, Figure **B**, the vertical displacement is negative if the initial point of consideration is the maximum height and the final point is the landing point just before the projectile hits the ground.

87. For problem **52**, the displacement is negative if the ball is coming down the inclined plane. Assume *x* and *y* directions are as shown.

88. For problem **56**, the vertical displacement between point B and the landing point is −25 m.

89. For problem **68**, the vertical displacement between points A and B is zero. They are at the same level.

90. For problem **68**, the vertical displacement of point A is positive with respect to the initial position of the launching point (the origin).

91. For problem **74**, the vertical displacement between the exit point of the inclined plane and the 12-m point is 9 m.

92. The magnitude of a velocity vector $(v_{fx}{}^2 + v_{fy}{}^2)^{1/2}$ continuously changes for a projectile as v_{fy}, the final vertical velocity component, changes in magnitude due to the action of acceleration due to gravity.

93. In most cases, the angle of the final velocity vector with respect to the horizontal at a time *t* is different from the projection angle α.

94. The angle of the velocity vector with respect to the horizontal ($tan\ \theta = v_{fy} / v_{fx}$) for a projectile changes as v_{fy} changes during the course of a projectile's trajectory.

95. Acceleration due to gravity is −10 m/s/s only if positive *y* is defined as vertically up.

96. If the (final) velocity vector is in the first quadrant (*x* and *y* components are positive), then the horizontal and vertical components can be equal in magnitude only if the angle of projection is 45 degrees with respect to the horizontal.

97. Oops! *sin* 30 = *cos* 60.

98. Oops! *sin* 60 = *cos* 30.

99. Oops! *sin* 15 = *cos* 75.

100. Oops! *sin* 28 = *cos* (90 − 28).

101. For a projectile, in the absence of gravity, the horizontal component of the velocity vector is a constant in time.

102. As an object climbs towards the maximum height, vertical velocity is "up" along $+y$ but acceleration is "down" along $-y$.

103. If velocity and acceleration components along the vertical have opposite signs, then we can conclude that the object is decelerating or slowing down.

104. If velocity and acceleration components along the vertical have the same sign, then we can conclude that the object is accelerating or speeding up or covering more distance per unit time in successive time intervals.

105. If a *velocity-time* graph is a straight line with a non-zero slope, then acceleration is a constant.

106. For a projectile, the square of the magnitude of a velocity vector $v^2 = (v_{fx}^2 + v_{fy}^2)$ also changes with time because v_{fy}, the final vertical velocity component, changes in magnitude due to the action of acceleration due to gravity.

107. The square of the magnitude of a velocity vector $v^2 = (v_{fx}^2 + v_{fy}^2)$ at maximum height for a projectile that is launched at an angle with respect to the horizontal reduces to $v^2 = (v_{fx}^2 + 0)$ as speed along the vertical at the maximum height is zero.

108. For problem **107**, for a projectile that is launched at an angle with respect to the horizontal, the square of the magnitude of a velocity vector $v^2 = (v_{fx}^2 + v_{fy}^2)$ at maximum height is a non-zero number.

109. If an object is thrown vertically up (that is the horizontal component of the velocity vector is zero), the square of the magnitude of the velocity vector $v^2 = (v_{fx}^2 + v_{fy}^2)$ at maximum height is $0 + 0 = 0$.

110. For two points A and B at the same height as in problem **68**, $v^2 = (v_{fx}^2 + v_{fy}^2)$ is the same because the square of a negative number (at B, v_{fy} is negative because it is on its way down) is a positive quantity.

111. For problem **36**, Figure **A**, at the launch point, the square of the magnitude of the velocity vector $v^2 = (v_{fx}^2 + v_{fy}^2)$ becomes $v^2 = (v_{0x}^2 + 0)$ as initial velocity along the vertical is 0 m/s.

112. The square of the magnitude of the velocity vector $v^2 = (v_{fx}^2 + v_{fy}^2)$ is the same for Figures D and E of problem **36**.

113. Two balls of different size are kicked at the same instant, with the same speed, and at the same angle with respect to the horizontal.

In the absence of air resistance, the two trajectories will be the same only if kicked at the same speed.

3

For problems **1-41**, choose the correct free body diagram (FBD). Assume g is the acceleration due to gravity. For problems **1-28**, all vectors are drawn to scale unless otherwise noted.

1. A computer monitor is resting on a table. Assume m is the mass of the monitor and N is the normal force.

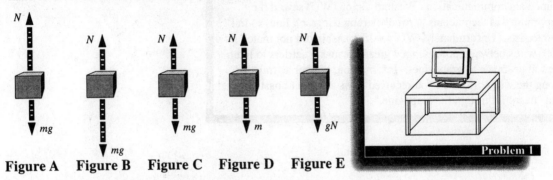

Figure A   **Figure B**   **Figure C**   **Figure D**   **Figure E**

2. A monitor rests on a two-dimensional table (assume two legs). M_m and M_T are the masses of the monitor and the table, respectively. N_1 and N_2 are the normal forces, respectively.

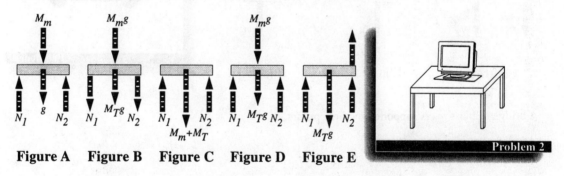

Figure A   **Figure B**   **Figure C**   **Figure D**   **Figure E**

3. A two-dimensional chair of mass m_C is sitting on the floor.

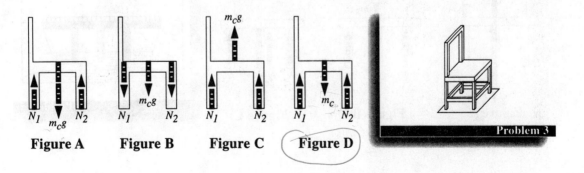

Figure A   **Figure B**   **Figure C**   **Figure D**

Pay No Attention to That Man Behind the Curtain
by Dan Barth

The phrase "I heard it through the grapevine" has its origins in the burgeoning communications industry. The rapidity of the westward movement underscored the need for clear and immediate communication. Western Union (WU) seized the opportunity as technicians hurriedly strung telegraph lines to link both coasts. Unfortunately, WU's workmanship did not match its zeal: wires between poles sagged greatly, causing settlers to liken their appearance to grapevines. Information, of course, traveled along the wires, so when you received news on either coast you had "heard it through the grapevine."

4.   A car of mass m_C rests on the ground.

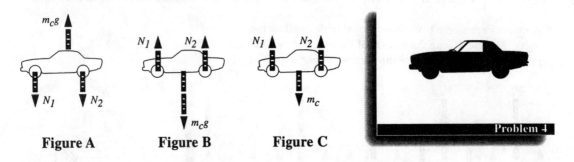

| **Figure A** | **Figure B** | **Figure C** | **Problem 4** |

5.   A bridge of mass m_B is supported by three pillars.

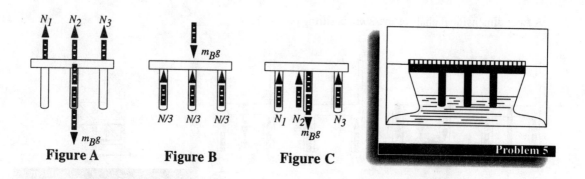

| **Figure A** | **Figure B** | **Figure C** | **Problem 5** |

6. A painting is hung from the metal rings as shown. Assume no strings.

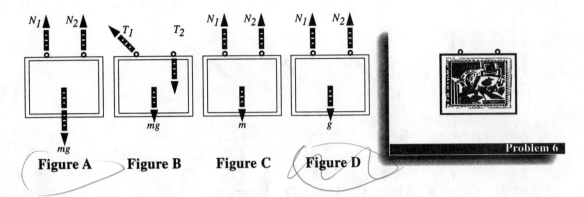

Figure A   Figure B   Figure C   Figure D

Problem 6

Common Physics Misconceptions
By Joseph Straley

It's a sad fact, but physics is based on a series of misconceptions about how the world really works. Your physics professor thinks in terms of these crazy ideas; you will need to learn what they are and to speak the physics language, or your physics professor will not be able to understand you.
For instance, physics professors think objects will move in a straight line at constant speed all by themselves.
In the real world, things move only when we push on them, as expressed by the equation $F = me$. However, in the magical world inside your professor's head, they coast forever, effortlessly. Explain to your professor that you were aimed directly at an A when you started the course and have been careful not to exert yourself since, so you will surely end with
an A, by Newton's First Law.

7. A fish on a fishing line is accelerating to the left.

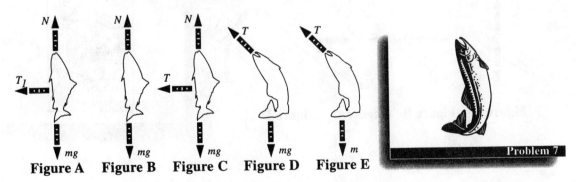

Figure A   Figure B   Figure C   Figure D   Figure E

Problem 7

8. A fish on a fishing line is accelerating to the right.

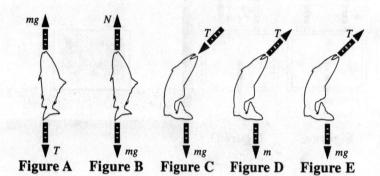

Figure A   **Figure B**   **Figure C**   **Figure D**   **Figure E**

9. A fish is held steady by means of a line.

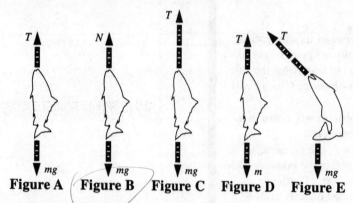

Figure A   **Figure B**   **Figure C**   **Figure D**   **Figure E**

10. A bat has just hit a ball.

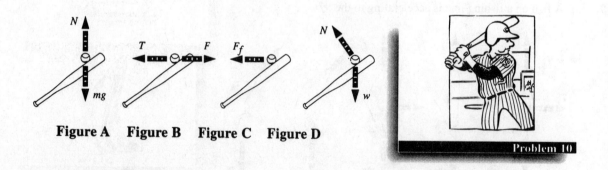

Figure A   **Figure B**   **Figure C**   **Figure D**

Problem 10

11. The ball has left the bat.

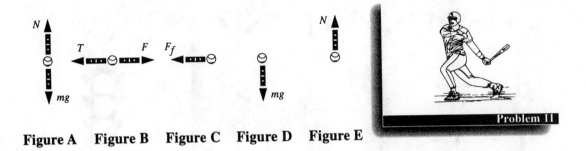

Figure A   Figure B   Figure C   Figure D   Figure E

12. A hockey puck is going straight on the ice.

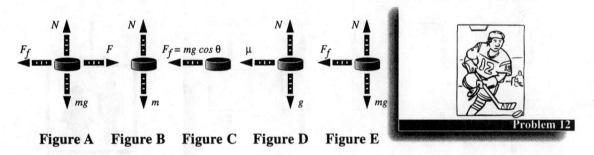

Figure A   Figure B   Figure C   Figure D   Figure E

13. A car is sliding to the right. Assume F_f is the force of friction, and μ is the coefficient of friction.

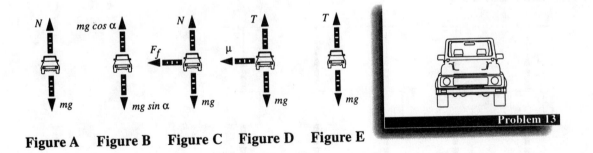

Figure A   Figure B   Figure C   Figure D   Figure E

14. A signal light is hung as shown. Assume the forces act on the junction.

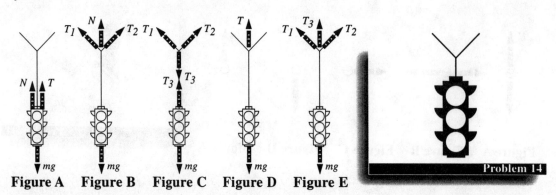

Figure A **Figure B** **Figure C** **Figure D** **Figure E**

Problem 14

15. An elevator is held at rest by a cable. Assume m is the total mass and T is the tension in the cable.

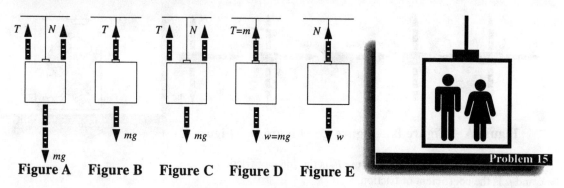

Figure A **Figure B** **Figure C** **Figure D** **Figure E**

Problem 15

For problems **16-19**, choose one of the following as the correct answer.

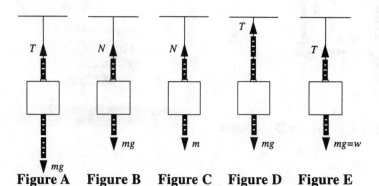

Figure A **Figure B** **Figure C** **Figure D** **Figure E**

16. The elevator is accelerating downward. A

17. The elevator is accelerating upward. D

18. The elevator is now going upward at constant velocity.

19. The elevator is moving downward at constant velocity. E

20. A roller coaster is on the inside of a loop. A

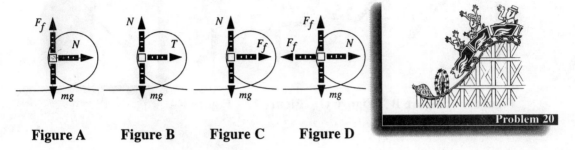

Figure A   **Figure B**   **Figure C**   **Figure D**

Problem 20

21. A roller coaster is at the position indicated.

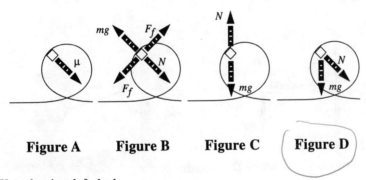

Figure A   **Figure B**   **Figure C**   **Figure D**

22. Water has just left the hose.

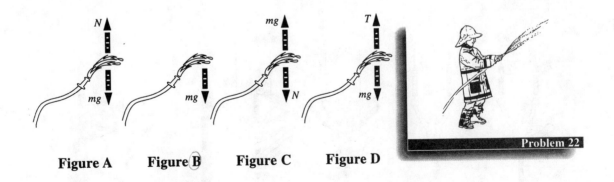

Figure A   **Figure B**   **Figure C**   **Figure D**

Problem 22

23. A skier is in free fall after jumping from the top of a cliff.

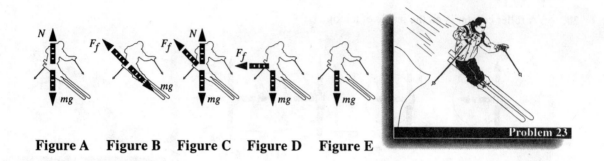

Figure A   **Figure B**   **Figure C**   **Figure D**   **Figure E**

24. A space shuttle is in free fall along $-y$ through the atmosphere. Assume non-zero air resistance, F_R, and that the space shuttle is accelerating vertically downward.

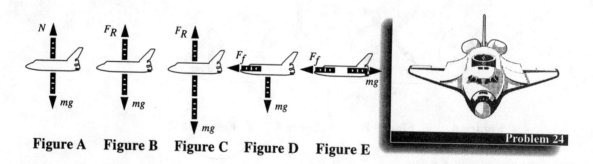

Figure A   **Figure B**   **Figure C**   **Figure D**   **Figure E**

25. A gymnast is hanging from the rings.

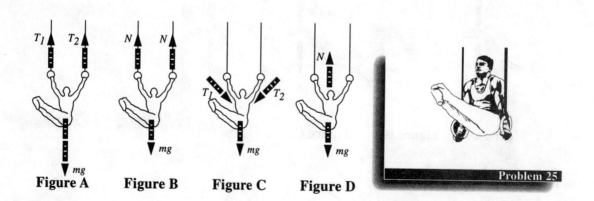

Figure A   **Figure B**   **Figure C**   **Figure D**

26. A banana is hanging from the ceiling.

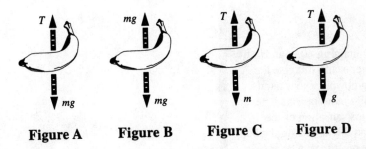

Figure A   **Figure B**   **Figure C**   **Figure D**

27. A pulley with a non-zero mass m is accelerating down. Please look at the figure to see m_1 and m_2.

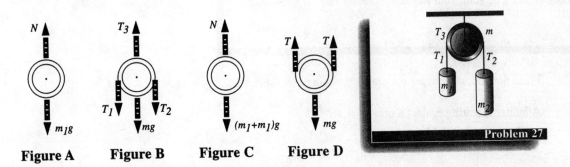

Figure A   **Figure B**   **Figure C**   **Figure D**

Problem 27

28. A football is in free fall in the air and is accelerating downwards. Assume non-zero air resistance, F_R. Same choice for forces as in problem **24**.

> **What You Feel is Not Necessarily What a Physics Professor Sees!**
> **by Joseph Straley**
>
> Suppose you are holding a box containing 40 pounds of gold(isn't that a nice idea?). You can rather easily imagine the sensation of the weight of the box pushing down on your hands, having at various times in your life held 40-pound objects – never gold, but perhaps fertilizer or potatoes. This sensation of the weight on your arms, which you can almost feel if you think about it, is completely irrelevant to your physics professor, who is only interested in the forces on the box. The force the box exerts on you is almost out of the picture – it is most nearly related to a force on the box (due to your hands) – in the upwards direction – equal and opposite to the one you had in mind – which combines with the gravitational force on the box to give no net force on the box – assuming that it is not accelerating. So where you feel the force of a box of gold, your physics professor sees ... nothing at all!

From now on, vectors are not drawn to scale unless specified.

29. A ladder is resting against a smooth wall.

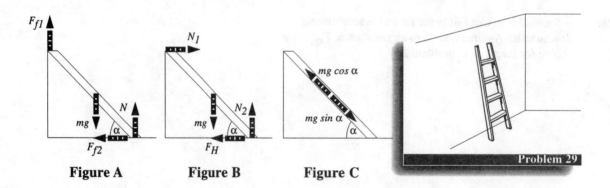

Figure A   Figure B   Figure C

Problem 29

30. A person of mass m_p stands on a ladder.

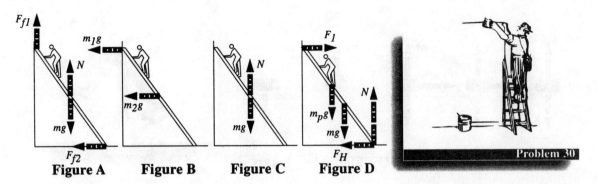

Figure A **Figure B** **Figure C** **Figure D**

Problem 30

31. An object hangs as shown from a nail.

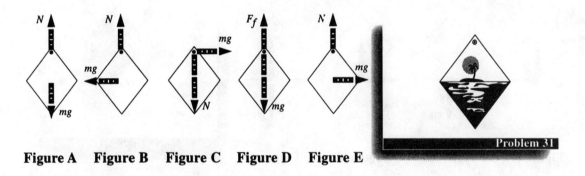

Figure A **Figure B** **Figure C** **Figure D** **Figure E**

Problem 31

32. A painting of mass m_p is held by two people (m_1 and m_2) standing on a platform that is supported by two ropes. Assume T_1 and T_2 are the tensions within the rope.

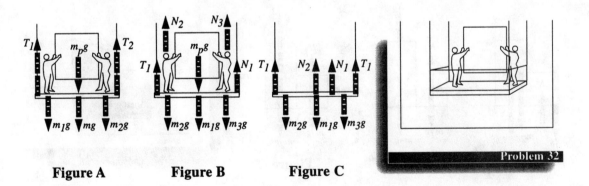

Figure A **Figure B** **Figure C**

Problem 32

33. A hockey puck is sliding to rest under friction on a level road.

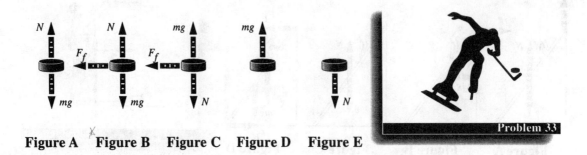

Figure A   **Figure B**   **Figure C**   **Figure D**   **Figure E**

34. A box sits ready to be lifted off the ground.

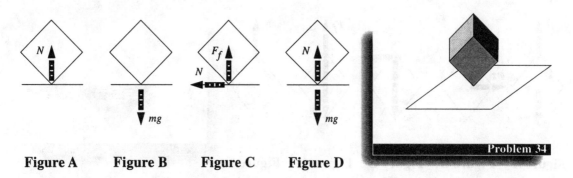

Figure A   **Figure B**   **Figure C**   **Figure D**

35. Two masses as shown as they are pushed to the right.

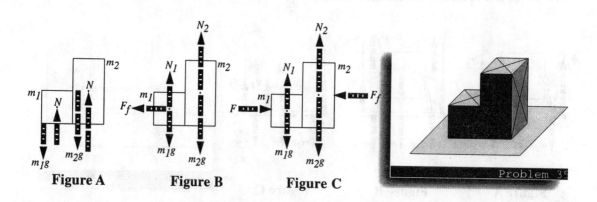

Figure A   **Figure B**   **Figure C**

36. A lady in a cart tries to pull herself up by pulling down on a rope.

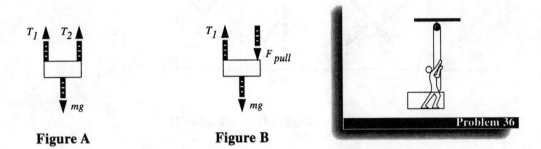

Figure A   **Figure B**   Problem 36

37. Two masses are hung as shown.

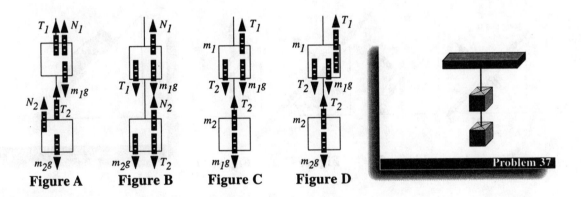

Figure A   **Figure B**   **Figure C**   **Figure D**   Problem 37

38. A mass is held on an inclined plane as shown.

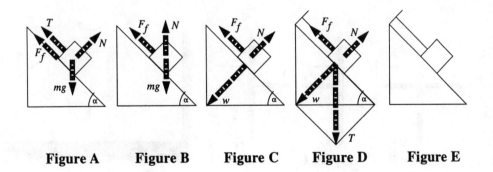

Figure A   **Figure B**   **Figure C**   **Figure D**   **Figure E**

39.   A mass is accelerating down on an inclined plane. Assume no friction.

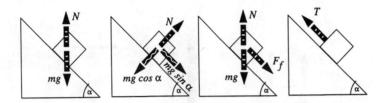

Figure A   **Figure B**   **Figure C**   **Figure D**

40.   A car (with the engine turned off) on an inclined plane is going down at constant speed.

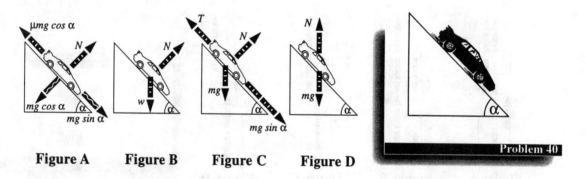

Figure A   **Figure B**   **Figure C**   **Figure D**

Problem 40

41.   An amusement park rotor ride (a cylindrical shell) rotates with a person held to the inside of the body of the cylinder.

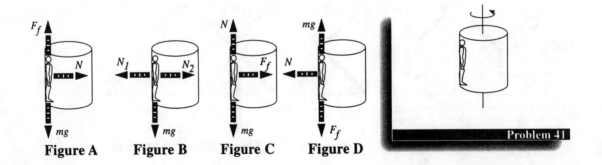

Figure A   **Figure B**   **Figure C**   **Figure D**

Problem 41

42. The two masses are situated as shown and the mass m_1 is moving in circles. Ignore friction. Choose the FBD for m_1.

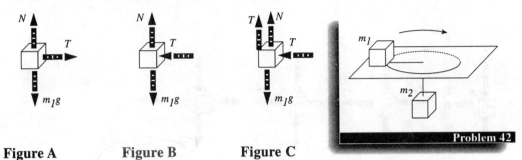

Figure A   **Figure B**   **Figure C**

43. Choose the FBD diagram of stationary mass m_2 in problem **42**. Assume vectors are drawn to scale.

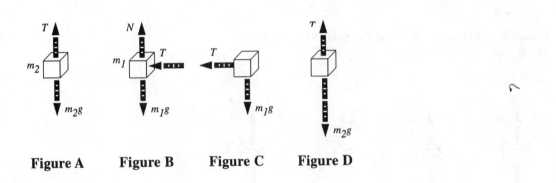

Figure A   **Figure B**   **Figure C**   **Figure D**

44. Two masses are going in circles as shown. Choose the FBD for the mass m_1.

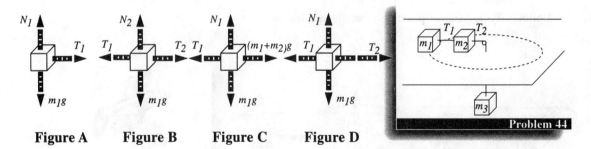

Figure A   **Figure B**   **Figure C**   **Figure D**

45. Draw the FBD for the mass m_2 in problem **44**. Assume vectors are drawn to scale.

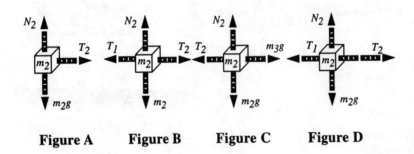

Figure A   **Figure B**   **Figure C**   **Figure D**

46. Choose the FBD for the mass m_3 of problem **44**. Assume steady case (equilibrium conditions).

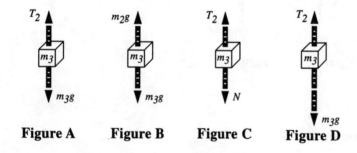

Figure A   **Figure B**   **Figure C**   **Figure D**

47. A pilot is flying a fighter plane horizontally upside down.

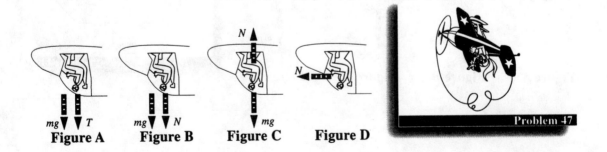

Figure A   **Figure B**   **Figure C**   **Figure D**

Problem 47

48. Choose the FBD for the same case as in problem **47**, but with the pilot sitting up and the plane right-side-up.

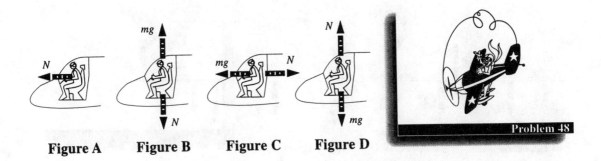

Figure A   **Figure B**   **Figure C**   **Figure D**

Problem 48

49. A person is carrying books. Choose the FBD for the books.

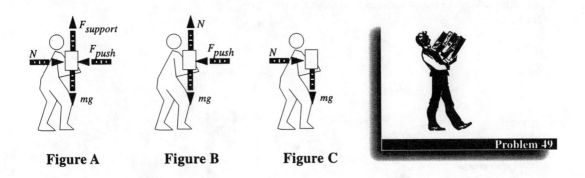

Figure A   **Figure B**   **Figure C**

Problem 49

50. Two people of unequal masses are sitting on a seesaw. Choose the FBD for the seesaw.

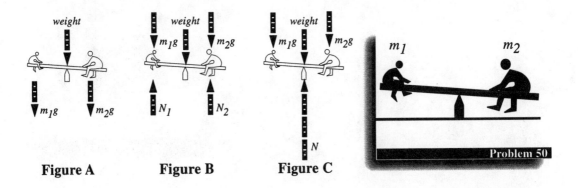

Figure A   **Figure B**   **Figure C**

Problem 50

51. A person is standing on the end of a diving board. Draw the FBD for the diving board.

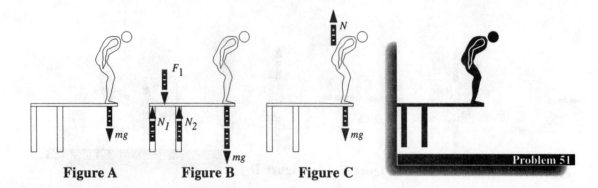

Figure A   **Figure B**   **Figure C**   Problem 51

52. A person is trying to pull a box to the right. The box is attached to a pulley on the left. Indicate the forces acting at the coupling where the ropes come together, but not the forces on the box.

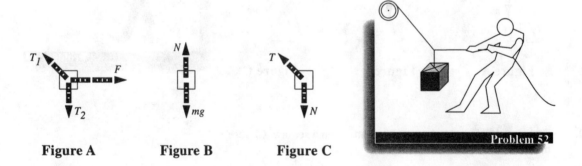

Figure A   **Figure B**   **Figure C**   Problem 52

53. A water skier is skiing while attached to the back of a boat. Assume non-zero friction and that he is ready to tip over. For convenience, assume any frictional force present to be horizontal. Choose the FBD for the skier.

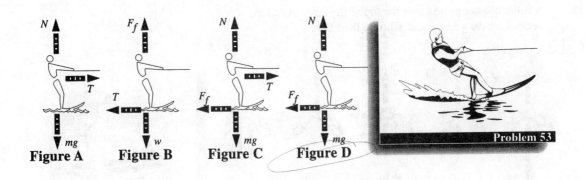

Figure A   **Figure B**   **Figure C**   **Figure D**

Problem 53

54. The skier in problem **53** lets go of the rope. Assume the same choices as above. Choose the FBD.

55. A bottle lies as shown. Choose the FBD.

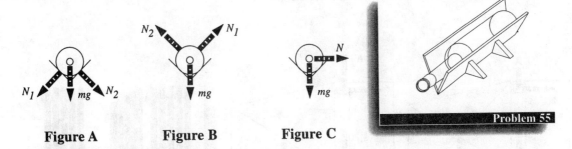

Figure A   **Figure B**   **Figure C**

Problem 55

For problems **56–58**, draw the FBD for the pendulum bob at the positions indicated. Choose one of the following choices as the correct answer.

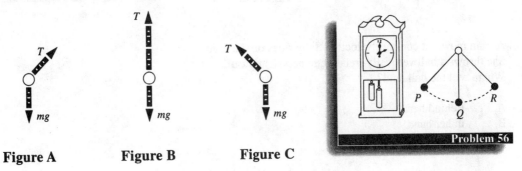

Figure A   **Figure B**   **Figure C**

Problem 56

56.   At the position P.

57.   At the position Q.

58.   At the position R.

59.   A roller coaster passes over the top of the rails. Assume vectors are drawn to scale. Choose the FBD.

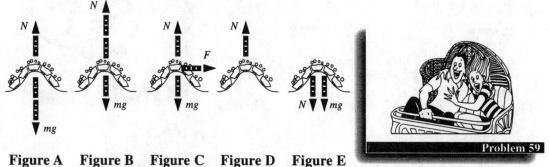

Figure A   **Figure B**   **Figure C**   **Figure D**   **Figure E**

60.   A gate is supported by two hinges. For convenience, ignore other forces on the door. Choose the FBD for the gate.

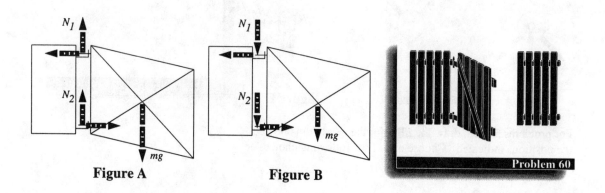

Figure A   **Figure B**

61.   A train moves at constant velocity. Sally rides on the train. She throws a ball vertically (relative to herself) upward. Where will the ball land?

A.   Behind her.
B.   In her hand.
C.   In front of her.

62.   Two bodies of different masses ($m_1 > m_2$) are dropped simultaneously from the same height to the ground. Which mass reaches the ground first if the air resistance (F_R) is the same on both the bodies?

     A. m_2.
     B. m_1.
     C. m_1 and m_2 will land simultaneously.

For problems **63–68**, Joe stands on a scale to weigh himself in an elevator. His mass is 60 kg. Choose from answers **A–E** for the scale reading in each situation. Assume g is 10 m/s^2.

     A.   480 N or 48 kg.
     B.   600 N or 60 kg.
     C.   0 N or 0 kg.
     D.   602 N or 60.2 kg.
     E.   720 N or 72 kg.

63.   The elevator is at rest.

64.   The elevator is rising at a constant acceleration of 2 m/s^2.

65.   The elevator is descending at a constant acceleration of 2 m/s^2.

66.   Whoops! The elevator is falling freely.

67.   The elevator is rising at a uniform speed of 2 m/s.

68.   The elevator is descending at a uniform speed of 2 m/s.

For problems **69–73**, a toy elevator of mass 0.1 kg is held by a massless cable. Calculate the tension in the cable for each situation, using answers **A–D**. Assume g is 10 m/s^2.

     A.   0.7 newton.
     B.   1.3 newtons.
     C.   1 newton.
     D.   0 newton.

69.   The elevator is at rest.

70.   The elevator is rising at a constant acceleration of 3 m/s^2.

71.   The elevator is descending at a constant acceleration of 3 m/s^2.

72. As the elevator is rising, it decelerates at 3 m/s² while preparing to stop.

73. As the elevator is descending, it decelerates at 3 m/s² while preparing to stop.

74. An elevator of mass m_e is held by a cable of mass m_c. What is the tension T_1 felt within the upper part of the cable at the instant it starts to move, if the elevator accelerates up at a constant acceleration of a m/s²?

    A. $\quad (m_e - m_c) \times (g + a).$
    B. $\quad (m_e - m_c) \times (g - a).$
    C. $\quad (m_e + m_c) \times (g + a).$
    D. $\quad (m_e + m_c) \times (g - a).$

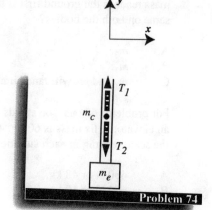

Problem 74

75. Two masses $m_1 = 0.2$ kg and $m_2 = 0.25$ kg are connected by a massless string over a massless pulley, and they are released from the same height. What is the acceleration of the system?

    A. $\quad$ 2.22 N.
    B. $\quad$ 2.22 m/s².
    C. $\quad$ 1.1 N.
    D. $\quad$ 1.1 m/s².

76. For problem **75**, what is the tension within the cable?

    A. $\quad$ 2.22 N.
    B. $\quad$ 2.22 m/s².
    C. $\quad$ 1.1 N.
    D. $\quad$ 1.1 m/s².

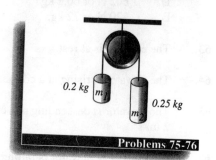

Problems 75-76

Two masses $m_1 = 2$ kg and $m_2 = 3$ kg are connected by a massless string over a massless pulley. The table and the pulley are frictionless.

77. What is the acceleration of the system?

    A. $\quad$ 2 m/s².
    B. $\quad$ 4 m/s².
    C. $\quad$ 6 m/s².
    D. $\quad$ 8 m/s².
    E. $\quad$ 5 m/s².

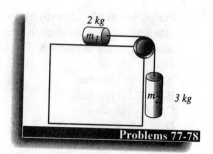

Problems 77-78

78. For problem **77**, what is the tension felt within the string?

    A.   6 N.
    B.   12 N.
    C.   24 N.
    D.   48 N.

For problems **79–84**, three masses are pulled to the right along the horizontal with a force $F = 20$ N. The masses are: $m_1 = 2$ kg, $m_2 = 3$ kg, $m_3 = 5$ kg. Assume that the table is frictionless and acceleration is uniform.

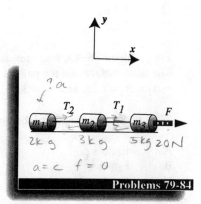

Problems 79-84

79. What is the tension T_1?

    A.   10 N.
    B.   20 N.
    C.   4 N.
    D.   15 N.

$\Sigma F_x = F_p - F_{T_1} = ma$

$20 - F_{T_1} = 5(a)$

80. What is the tension T_2?

    A.   10 N.
    B.   20 N.
    C.   4 N.
    D.   15 N.

$F_{m_3} = 20N$

$F_{sm_3} = -20N$

$F + 20N = m_3 a_x$

$F + 20N = 10kg (m_1 + m_2 + m_3) a_x$

$a_x = 2$

81. What is the acceleration of m_1?

    A. 10 m/s^2.
    B. 35 m/s^2.
    C. 4 m/s^2.
    D. 2 m/s^2.

82. The acceleration of m_2 is not the same as that of m_1.

    A.   True.
    B.   False.

83. If m_1 started from rest, what is its speed after one second?

    A.   10 m/s.
    B.   35 m/s.
    C.   4 m/s.
    D.   2 m/s.

84.   If m_2 started from rest, what distance has it moved after
       3 seconds? Assume constant acceleration.

       A.   10 m.
       B.   3.5 m.
       C.   4.8 m.
       D.   9 m.

       For problems **85-89**, four masses are held by strings as shown.
       The table surface and the pulleys are frictionless. Assume m_1
       $= m_2 = m_3 = 4$ kg, and $m_4 = 3$ kg.

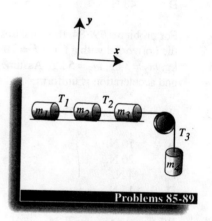

Problems 85-89

85.   What is the acceleration of the system?

       A. 10 m/s^2.
       B. 5 m/s^2.
       C. 4 m/s^2.
       D. 2 m/s^2.

86.   What is the speed of m_4 after 2 seconds, if it started from rest?

       A.   10 m/s.
       B.   35 m/s.
       C.   4 m/s.
       D.   2 m/s.

       For problems **87-89**, choose one of the following answers.

       A.   30 N.
       B.   8 N.
       C.   16 N.
       D.   12 N.
       E.   24 N.

87.   What is the tension T_3?

88.   What is the tension T_2?

89.   What is the tension T_1?

For problems **90-92**, an object of mass m descends an inclined plane of angle α. Assume no friction.

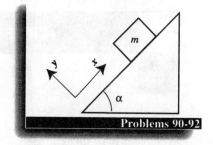

Problems 90-92

90. What is the acceleration?

    A. g.
    B. $g \times sin\ \alpha$.
    C. $g \times cos\ \alpha$.
    D. $g \times tan\ \alpha$.

For problems **91-92**, choose one of the following choices.

    A. $m \times g$.
    B. $m \times g \times sin\ \alpha$.
    C. $m \times g \times cos\ \alpha$.
    D. $m \times g \times tan\ \alpha$.

91. What is the component of the weight along the inclined plane?

92. What is the component of the weight perpendicular to the inclined plane?

93. Assume that the plane and the pulley are smooth. Find the acceleration of the system if $m_1 = 2$ kg, $m_2 = 3$ kg, and $\alpha = 30°$.

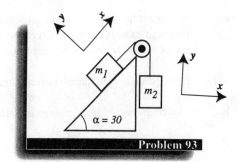

Problem 93

    A. $2\ \text{m/s}^2$.
    B. $4\ \text{m/s}^2$.
    C. $6\ \text{m/s}^2$.
    D. $8\ \text{m/s}^2$.

94. For problem **93**, what is the tension felt within the string?

    A.   6 N.
    B.   12 N.
    C.   18 N.
    D.   24 N.

95. A mass of 25 kg is pulled to the right by a rope of mass 2 kg along a frictionless horizontal plane by a horizontal force of 50 N. What is the force exerted on the mass?

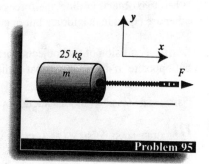

Problem 95

    A.   50 N.
    B.   46.3 N.
    C.   53.7 N.
    D.   3.7 N.

96. Net force is mass times acceleration.

97. In problem **1**, normal force is the perpendicular force exerted by the table surface on the computer monitor.

98. Weight always acts vertically down.

99. If two forces acting on the same object are equal in magnitude but opposite in directions, then the net force on the object is zero.

100. An object is moving at constant velocity. Suddenly two equal forces in exactly opposite directions act on the object. Under the new circumstances, the net force is still zero.

101. If net force is zero, acceleration is zero.

102. If acceleration is zero, change in velocity is also zero.

103. If acceleration is zero, velocity is a constant.

104. An object is moving in a circle at a uniform speed, and it continuously changes direction. Because velocity has a magnitude and direction, if one of the two changes (in this case, direction), then velocity also changes.

105. When two objects collide, the forces exerted by each on the other are equal in magnitude but opposite in direction.

106. During a collision of two objects, forces are equal in amount but opposite in direction. If that is the case, we can write:

$$m_{\text{SMALLER MASS}} \times a_{\text{BIGGER ACCELERATION}} =$$

$$m_{\text{BIGGER MASS}} \times a_{\text{SMALLER ACCELERATION}}.$$

107. When you fire a bullet, the force exerted by the gun on the bullet is equal in magnitude to the force exerted by the bullet on the gun. We can then write:

m(SMALLER) MASS OF THE BULLET ×

a(BIGGER) ACCELERATION OF THE BULLET =

m(BIGGER) MASS OF THE GUN ×

a(SMALLER) ACCELERATION OF THE GUN.

108. When you drop a pen, the force exerted by the pen on the earth is equal in magnitude to the force exerted by the earth on the pen.

M(SMALLER) MASS OF THE PEN ×

a(BIGGER) ACCELERATION OF THE PEN =

m(BIGGER) MASS OF THE EARTH ×

a(SMALLER) ACCELERATION OF THE EARTH.

109. A mother (bigger mass) pushes against her child on a frictionless ice pond. The force exerted by the child on the mother is equal in magnitude to the force exerted by the mother on the child.

m(SMALLER) MASS OF THE CHILD ×

a(BIGGER) ACCELERATION OF THE CHILD =

m(BIGGER) MASS OF THE MOTHER ×

a(SMALLER) ACCELERATION OF THE MOTHER.

110. A small car collides against a Mack truck.
The force exerted by the car on the Mack truck is equal in magnitude to the force exerted by the mack truck on the car.

m(SMALLER) MASS OF THE CAR ×

a(BIGGER) ACCELERATION OF THE CAR =

m(BIGGER) MASS OF THE MACK TRUCK ×

a(SMALLER) ACCELERATION OF THE MACK TRUCK.

111. For the above-mentioned cases, bigger acceleration means bigger change in velocity.

112. If the monitor is placed off-center for problem **2**, the normal forces N_1 and N_2 exerted by the floor on the table will not be equal.

113. For problem **3**, the weight of the chair acts vertically down, but the normal force exerted by the floor on the chair is vertically up.

114. For problem **4**, the sum of the normal forces exerted by the ground on the car (at the front and back wheels) is equal to the weight of everything inside the car plus the car's weight.

115. An object accelerates to the left because there is a net unbalanced force in that direction.

117. For problem **10**, the normal force exerted by the bat on the ball is perpendicular to the surface at the point of contact.

118. If there is a net force, there is a non-zero acceleration.

119. If there is a non-zero acceleration, velocity changes.

120. When there is a change in velocity, its magnitude or direction or both magnitude and direction can change.

121. For problem **12**, the weight of the hockey-puck acts vertically down, the normal force exerted by the surface on the puck acts vertically up perpendicular to the surface, and the force of friction is opposite to the direction of motion.

122. In a certain direction, with two forces acting in opposite directions on the same object, we can write (using their magnitudes):
Bigger Force – Smaller Force = Net Force.

123. We do not always need two forces to create a net force. In general, there can be one, two, or more forces that can act on an object in a certain direction.

124. An object can move without any forces acting on it.

125. An object can move with forces acting on it.

126. An object can move at constant velocity without any forces acting on it.

127. An object can move at constant velocity with forces acting on it.

128. An object can move at constant acceleration under the action of a single constant force.

129. An object can move at constant acceleration under the action of two forces.

130. An object moves along +*x*. That does not mean that the net force is along +*x*.

131. An object is thrown vertically up. While climbing to the maximum height, its velocity is along +*y* even though net force on it is along −*y* in the vertically down direction due to the force of gravity.

132. A hockey puck moves along +*x* on a rough surface. The net force is along −*x*.

133. When net force and velocity are in the same direction, the object speeds up.

134. An object is thrown vertically up. While coming down from the maximum height, its velocity is along −*y*. Net force on it is also along −*y* in the vertically down direction due to the force of gravity. The object speeds up.

135. If two balls of the same mass bounce horizontally off a vertical wall, the one that bounces with a bigger speed imparts more force to the wall. Assume the same time of contact.

136. After you hit a hockey puck, the hitting force does not remain with the puck once the hockey stick loses contact with the puck.

137. In problem **23**, the skier does not experience a normal force because this person is no longer in contact with a surface.

138. Normal force can be equal to the weight of an object.

139. Normal force can be greater than the weight of an object.

140. Normal force can be less than the weight of an object.

141. For problem **29**; the ladder has a tendency to slip (or slide) to the right.

142. For problem **29**; the ladder has a tendency to slip (or slide) to the right, so the force of friction has to be to the left.

143. For problem **29**; if the ladder is not sliding, the static force of friction is the force that acts to the left on the ladder.

144. For problem **29**; normal force has to act at the point where the surface is in contact with the ladder.

145. For problem **29**; the magnitude of the force exerted by the wall on the ladder is equal to the magnitude of the force exerted by the ladder on the wall.

146. For problem **29**; to draw a Free Body Diagram (FBD) of the ladder, we exclude the force exerted by the ladder on the wall but include the force exerted by the wall on the ladder.

147. For problem **29**; if the wall is frictionless, the ladder is pushed by the wall horizontally to the right.

148. As a general statement, for problems involving a ladder and a wall, the wall does not always have to push the ladder horizontally.

149. A region of non-uniform forces acting on a moving object can be due to patches of ice, snow, and water if the object comes in contact with them.

150. In problem **41**, we should ignore the force exerted by the person on the wall when we choose the correct FBD for the person.

4

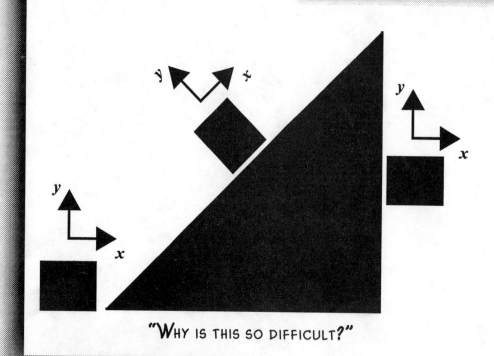

"WHY IS THIS SO DIFFICULT?"

Answer the following problems related to the equation of motion.

1. A body of mass m is at rest on a table with no external forces acting on it. What balances the weight?

    A. Normal force, N.
    B. Mass.

2. For problem **1**, what is the static frictional force if the coefficient of friction is μ?

    A. μ.
    B. $F_f = \mu N = \mu mg$.
    C. μm.
    D. Zero.

    For problems **3–4**, a body of mass m is at rest on a horizontal table. Assume a downward force of 10 N in addition to its weight acts on the mass.

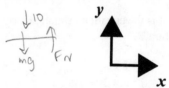

3. Balance the force components along y.

    A. Normal force = weight.
    B. Normal force + downward force = weight.
    C. Weight + downward force = normal force.

4. In problem **3**, what is the static frictional force?

    A. μ.
    B. $\mu N = \mu (w + 10)$.
    C. $\mu N = \mu (w - 10)$.
    D. Zero.

5. An object of mass m slides to rest on a horizontal road extending along the horizontal x while sliding to the right at constant deceleration. What is the net force along y? Assume y is along the vertical.

    A. Zero.
    B. Non-zero.

6. In problem **5**, what is the net force along x?

    A.   Zero.
    B.   Non-zero.

7. For the object in problem **5**, what is the equation of motion along x?

    A.   Force of friction = normal force = ma.
    B.   Force of friction = $\mu \times$ normal force = ma.
    C.   Net force = 0.

8. For the object in problem **5**, what is the equation of motion along y? Assume y to be along the vertical.

    A.   Normal force = weight. $\Rightarrow$ Net force = 0.
    B.   Net force is not zero.

For problems **9–11**, a body of mass m at rest is acted on by a 15-N force as shown in the figure.

9. What is the equation of motion along x when there is no friction?

    A. $15 = ma_x$.
    B. $15 \cos 30 = ma_x$.
    C. $15 \sin 30 = ma_x$.

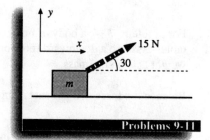

Problems 9–11

10. What is the equation of motion along x when friction is present? Coefficient of friction = μ.

    A. $15 = ma$.
    B. $\mu \times (15 \sin 30 + N) = ma_x$.
    C. $\mu N = ma. \Rightarrow \mu \times (mg + 15 \cos 30) = ma_x$.
    D. $\mu N = ma. \Rightarrow \mu \times (mg - 15 \cos 30) = ma_x$.
    E. $15 \cos 30 - \mu \times (mg - 15 \sin 30) = ma_x$.

11. What is the equation of motion along y (vertical)?

    A. $N = mg$.
    B. $N + (15 \sin 30) = mg$.
    C. $N + (15 \cos 30) = mg$.

[handwritten:]
$15 \sin 30°$
$\downarrow Fg$

$\Sigma Fy = ma$
$FN - Fg + Fp = 0$
$FN + 15 \sin 30 = mg$

12. A 15-N force acts (see figure) on an object of mass m, along the horizontal direction $+x$. What is the equation of motion along x? Assume no friction.

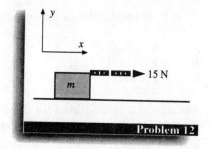

Problem 12

A. $15 = ma_x$.
B. $(15 + mg) = ma_x$.

13. Once we know acceleration (or for that matter deceleration), can we use the equation of kinematics, $v_f = v_0 + at$?

A. Yes.
B. No.

14. Once we know acceleration (or for that matter deceleration), can we use the equation of kinematics, $v^2 = v_0^2 + 2a\Delta x$?

A. Yes.
B. No.

15. Once we know acceleration (or for that matter deceleration), can we use the equation of kinematics, $\Delta x = v_0 t + (1/2) at^2$?

A. Yes.
B. No.

For problems **16–21**, a body of mass m is released from rest on an inclined plane. Write the equation of motion along x for the following conditions.

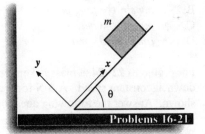

Problems 16–21

16. The mass m is accelerating down the inclined plane when released from rest. Assume no friction.

A. Net force along x is zero.
B. $mg \times \sin\theta = ma$.

17. The mass m is accelerating down after moving a distance, d. Assume constant acceleration and a smooth inclined plane.

A. $mg \times \sin\theta = mg \times \cos\theta$.
B. $mg \times \sin\theta = ma$.
C. $mg \times \cos\theta = ma$.

18.   What is the acceleration of the mass in problem **17**?

A. $g \times sin\ \theta$.
B. $g \times cos\ \theta$.
C. $g \times tan\ \theta$.

19.   The mass m is coming down but slowing at a constant
deceleration. Assume non-zero friction and that F_f is the force
of friction.

A. $mg\ sin\ \theta = mg\ cos\ \theta$.
B. $(mg\ sin\ \theta)\ -\ F_f = ma$.
C. $(mg\ cos\ \theta)\ -\ F_f = ma_x$.

20.   What is the force of friction on an inclined plane when there
are no external forces acting on the mass if the mass is sliding?

A. $F_f = \mu N = \mu mg\ sin\ \theta$.
B. $F_f = \mu N = \mu mg\ cos\ \theta$.

21.   If the mass is coming down the inclined plane at a constant
speed, what is the equation of motion? Assume non-zero
friction.

A. $mg\ sin\ \theta = mg\ cos\ \theta$.
B. $mg\ sin\ \theta = \mu mg\ cos\ \theta$.
C. $mg\ sin\ \theta = ma$.
D. $\mu mg\ cos\ \theta = ma$.

For problems **22–24**, a mass m kept on an inclined plane slides
down at constant speed. A 5-N force acts on the mass as
shown. Answer the following questions.

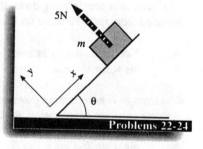

Problems 22-24

22.   What is the equation of motion along y?

A. $mg\ sin\ \theta = 5$.
B. $(N + 5) = mg\ cos\ \theta$.
C. $(N + 5) = mg\ sin\ \theta$.
D. $(N - 5) = mg\ sin\ \theta$.

23.   What is the force of friction?

A. $F_f = \mu mg$.
B. $F_f = \mu mg\ cos\ \theta$.
C. $F_f = \mu \times (mg\ cos\ \theta - 5)$.
D. $F_f = \mu \times (mg\ sin\ \theta - 5)$.

Meatballs
by Premy Augustus

1 lb (or more) of ground turkey, 1 onion, *minced fine,* 1/4 teaspoon oregano, ½ to 1 cup flavored bread crumbs, ½ to 1/4 cup of grated parmesan cheese, salt and black pepper to taste.

Mix the above items and form small meatballs in a baking tray sprayed with non-sticky oil (Pam). Add 1/4 cup of vegetable oil to the tray and bake at 375° for half an hour. Turn the meatballs over and bake for 15 more minutes. You can substitute beef if you like, but then add an egg white.

24. What is the equation of motion along x? Assume non-zero friction.

   A. $\mu \times (mg \cos \theta - 5) = mg \sin \theta$.
   B. $\mu \times (mg \cos \theta + 5) = mg \sin \theta$.
   C. $\mu mg \cos \theta = mg \sin \theta$.

For problems **25-29**, two masses m_1 and m_2 $(m_1 \gg m_2)$ are tied by a string of tension T and released from rest on a smooth inclined plane.

25. What is the equation of motion for the mass m_1? Assume no friction.

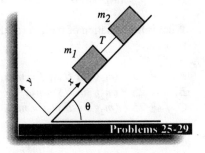

   A. $(m_1 g \sin \theta) - m_2 = m_2 a$.
   B. $(m_1 g \sin \theta) - T = m_1 a$.
   C. $(m_1 g \sin \theta) - T = (m_1 + m_2)a$.
   D. $T = (m_1 + m_2)a$.

26. What is the equation of motion for the mass m_2? Assume no friction.

   A. $T - m_2 g = m_2 a$.
   B. $T - (m_2 g \sin \theta) = m_2 a$.
   C. $T + (m_2 g \sin \theta) = m_2 a$.

27. What are the normal forces for masses m_1 and m_2?

   A. $m_1 g \cos \theta$ and $m_2 g \cos \theta$, respectively.
   B. $m_1 g \sin \theta$ and $m_2 g \sin \theta$, respectively.
   C. $m_1 g \cos \theta$ and $m_2 g \sin \theta$, respectively.

28. What is the equation of motion for mass m_1? Assume non-zero friction.

   A. $(m_1 g \sin \theta) - T = m_1 a$.
   B. $(m_2 g \sin \theta) - T = m_1 a$.
   C. $(m_1 + m_2) \times g \sin \theta = m_1 a$.
   D. $(m_1 g \sin \theta) - (T + F_f) = m_1 a$.

29. What is the equation of motion for the mass m_2? Assume non-zero friction.

   A. $(m_2 g \sin \theta) - T = m_2 a$.
   B. $(\mu m_2 g \cos \theta) - T - m_2 g = m_2 a$.
   C. $T + (m_2 g \sin \theta) - (\mu m_2 g \cos \theta) = m_2 a$.

For problems 30–33, two masses m_1 and m_2 are tied by a string and pulled up along the inclined plane at constant acceleration.

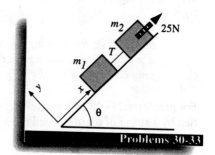

Problems 30-33

30. What is the equation of motion for the mass m_1? Assume no friction.

   A. $T - (m_1 g \sin \theta) = m_1 a$.
   B. $T - [(m_1 + m_2) \times g \sin \theta] = m_1 a$.
   C. $T = m_1 a$.
   D. $T - (m_1 g \sin \theta) - 25 = m_1 a$.

31. What is the equation of motion for the mass m_2? Assume no friction.

   A. $25 - (m_2 g \sin \theta) + T = m_2 a$.
   B. $25 - (m_2 g \sin \theta) = m_2 a$.
   C. $25 - (m_2 g \sin \theta) - T = m_2 a$.

32. What are the normal forces for the masses m_1 and m_2?

   A. $N_1 = m_1 g \cos \theta$ and $N_2 = m_2 g \cos \theta$.
   B. $N_1 = m_1 g \sin \theta$ and $N_2 = m_2 g \sin \theta$.
   C. $N_1 = m_1 g$ and $N_2 = m_2 g$.

33. Is the acceleration of mass m_1 the same as that of m_2?

   A. Yes.
   B. No.

For problems **34–37**, two masses m_1 and m_2 ($m_1 \gg m_2$) are connected by a massless rope of tension T, and released from rest.

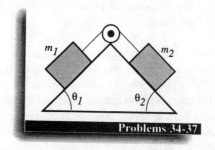

34. What is the equation of motion for mass m_1 if there is no friction?

    A. $m_1 g \sin\theta_1 = m_1 a$.
    B. $(m_1 g \sin\theta_1) - T = m_1 a$.
    C. $m_1 g \sin\theta_1 = (m_1 + m_2)a$.
    D. $(m_1 g \sin\theta_2) - T = m_1 a$.

35. What is the equation of motion for the mass m_2 if there is no friction?

    A. $T = m_2 a$.
    B. $T = (m_1 + m_2)a$.
    C. $T - (m_2 \sin\theta_2) = m_2 a$.
    D. $T - (m_2 g \sin\theta_2) = m_2 a$.
    E. $T - (m_2 g \sin\theta_1) = m_2 a$.

36. What is the equation of motion for the mass m_1 if friction is non-zero (a rough surface!)? Assume the same coefficient of friction μ for the two surfaces.

    A. $T - (m_1 g \sin\theta_1) - (\mu m_1 g \cos\theta_1) = m_1 a$.
    B. $(m_1 g \sin\theta_1) - T - (\mu m_1 g \cos\theta_1) = m_1 a$.
    C. $T = m_1 a$.
    D. $T = (m_1 + m_2)a$.

37. What is the equation of motion for the mass m_2 if friction is non-zero?

    A. $T - m_2 g = m_2 a$.
    B. $T - m_2 g \sin\theta_2 + \mu m_2 g \cos\theta_2 = m_2 a$.
    C. $T - m_2 g \sin\theta_2 - \mu m_2 g \cos\theta_2 = m_2 a$.

For problems **38–45**, three masses m_1, m_2, and m_3 are connected by cords of tensions T_1 and T_2 as shown in the figure. Assume $m_1 \gg (m_2 + m_3)$. Neglect friction.

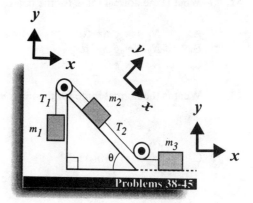

38. What is the equation of motion for the mass m_1?

    A. $m_1 g \sin\theta_1 - T_1 = m_1 a$.
    B. $T_1 - m_1 g \sin\theta = m_1 a$.
    C. $T_1 = m_1 a$.
    D. $m_1 g - T_1 = m_1 a$.

Pay No Attention to That Man Behind the Curtains
by Dan Barth

During the plague years of 1665 and 1666 Sir Isaac Newton laid the foundation for much of his life's work; critical to his research was his invention of fluxions – what we now call calculus. However, Newton did not share his discovery with the public, and after 1676 a dispute arose with the German mathematician, Gottfried Wilhelm Leibniz, over whose name should be associated with the origination of calculus. Later Newton pressed the issue to be decided by a "neutral" court, the Royal Society, of which he had been the President since 1703. Needless to say, Newton's primacy was upheld.

39. What is the equation of motion for the mass m_2?

    A. $T_2 + (m_2 g \sin \theta) - T_1 = m_2 a$.
    B. $T_1 - T_2 + (m_2 g \sin \theta) = m_2 a$.
    C. $T_1 + T_2 = m_2 g$.
    D. $T_1 - T_2 - (m_2 g \sin \theta) = m_2 a$.

40. What is the equation of motion for the mass m_3?

    A. $T_2 = m_3 a$.
    B. $T_2 = m_3 g$.
    C. $m_3 g = m_3 a$.

You Can Do It!

41. What is the normal force for the mass m_1? Assume no contact.

    A.   Zero.
    B.   Non-zero.

42. What is the normal force for the mass m_2?

    A. $N_2 = m_2 g \sin \theta$.
    B. $N_2 = m_2 g \cos \theta$.
    C. $N_2 = m_2 g$.

43. What is the normal force for the mass m_3?

    A. $N_3 = T_3$.
    B. $N_3 = m_3 g$.
    C. $N_3 = (m_2 + m_3) \times g$.

44. In problem **39**, what is the equation of motion for the mass m_2 when friction is non-zero?

    A. $T_2 + (m_2g \sin \theta) - T_1 = m_2a$.
    B. $\mu m_2g \cos \theta = m_2a$.
    C. $T_1 - (m_2g \sin \theta) - (\mu m_2g \cos \theta) = m_2a$.
    D. $T_1 - T_2 - (m_2g \sin \theta) - (\mu m_2g \cos \theta) = m_2a$.

45. In problem **40**, what is the equation of motion in the presence of friction?

    A. $T_2 - \mu m_3g = m_3a$.
    B. $\mu m_3g = m_3a$.
    C. $T_2 = m_3a$.

46. Two masses m_1 and m_2 ($m_1 >> m_2$) are connected by a massless rope as shown and released from rest. Assume no friction. What is the equation of motion for the mass m_1?

    A. $m_1g - T = m_1a$.
    B. $m_1g - T = (m_1 + m_2)a$.
    C. $m_1g = m_1a$.

47. In problem **46**, what is the equation of motion for the mass m_2?

    A. $m_2g - T = m_2a$.
    B. $m_2g - T = (m_1 + m_2)a$.
    C. $m_2g = m_2a$.
    D. $T - m_2g = m_2a$.

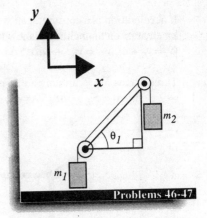

Problems 46-47

48. A person hangs on for dear life as shown by a rope tied to a stationary car. What should be the tension T in the rope if it is not to break?

    A.   Tension, $T >> mg$, the weight.
    B. $T << mg$.

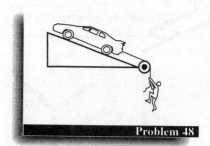

Problem 48

49. Two masses m_1 and m_2 ($m_1 >> m_2$) are acted on by a constant force (equal forces) to the right. Which one has a larger acceleration?

    A. m_1.
    B. m_2.
    C.   Both m_1 and m_2 have the same acceleration.

50. Two masses m_1 and m_2 ($m_1 \gg m_2$) slide down an inclined plane. If there is no friction, which one has the larger acceleration?

A. m_1.
B. m_2.
C. m_1 and m_2 have the same acceleration.

For problems **51-63**, choose one of the following answers.

A. Yes.
B. No.

51. If acceleration is a constant, can we use the equations of kinematics on an inclined plane when there is friction?
$[v_f = v_0 + at; \ v_f{}^2 = v_0{}^2 + 2a\Delta x; \ \Delta x = v_0 t + (1/2)at^2.]$

52. Three forces F_H, T, and mg are in equilibrium, as shown.
$\Rightarrow T / \sin 90 = F_H / \sin 120 = mg / \sin 150.$

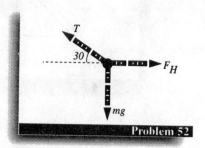

Problem 52

53. T_1, T_2, and mg are in equilibrium as shown.
$\Rightarrow T_1 / \sin 30 = T_2 / \sin 35.$

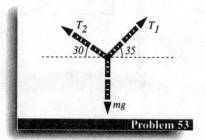

Problem 53

54. Refer to the figure of problem **53**. Assume equilibrium conditions. Is $T_1 / (90 + 30) = T_2 / (90 + 35) = mg / (180 - 30 - 35)$?

55. Three forces F_H, T_1, and mg are in equilibrium as shown.
$\Rightarrow F_H \, / \, sin \, (90 + 20) = T_1 \, / \, sin \, 90$.

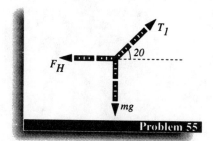

Problem 55

56. Masses m_1 and m_2 are at rest in equilibrium as shown.
$\Rightarrow T_2 \, / \, sin \, 135 = T_1 \, / \, sin \, 120 = m_2 g \, / \, [sin \, (180 - 30 - 45)]$.

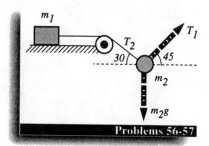

Problems 56-57

57. For problem **56**, can we write $\mu m_1 g \, / \, [sin \, 135] = m_2 g \, / \, [sin \, (180 - 30 - 45)] = T_1 \, / \, [sin \, 120]$.

Please refer to the figure of problem **58** for problems **58-63**.

58. In the figure, m_1, m_2, and m_3 are at rest. Can we then say that $T_3 = m_3 g$?

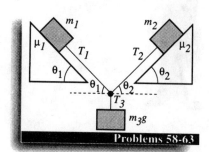

Problems 58-63

59. Can we say that $T_1 = \mu_1 m_1 g \cos \theta_1$?

60. Is $T_2 = \mu m_2 g \cos \theta_2$ for the mass m_2?

61. Is $T_1 / [\sin (90 + \theta_2)] = T_2 / [\sin (90 + \theta_1)] = m_3 g / [\sin (180 - \theta_1 - \theta_2)]$?

62. Is $(\mu m_1 g \cos \theta_1) / [\sin (90 + \theta_2)] = (\mu m_2 g \cos \theta_2) / [\sin (90 + \theta_2)]$?

63. Is $[m_3 g / \sin (180 - \theta_1 - \theta_2)] = T_2 / \sin \theta_1$?

64. When your car is on top of a hill, it is very easy to ride down (*even with the engine off – very dangerous: power brakes/steering don't work, steering may lock ...*) because there is a "free" pull on the car, which is actually the component of the weight acting parallel to the inclined plane.

65. If your car is at the base of a steep hill with the engine off, nothing happens, because there is no force available to take the car up the inclined plane.

66. A good design for a shopping center is to place the grocery store on a hill. That way, you as a consumer do not have to push the loaded grocery cart uphill or across a rough horizontal pavement.
[For the time being, assume empties are pulled uphill by motorized carts.]

67. On a modern road, the center portion is raised, so that when it rains, water drains to the gutters on the sides. Here the component of the weight pulls the water molecules to the base of the inclined plane.

68. Most of the roofs in this country are inclined planes because the component of the weight helps clear unwanted stuff (dust, dirt, snow, water, dry leaves ...) off the roof.

69. The steeper the roof, the greater the magnitude of the component of the weight acting parallel to the inclined plane.

70. The magnitude of the component of the weight acting parallel to an inclined plane = $m\,g\,\sin\theta$, where m is the mass in kg, g is the magnitude of the acceleration due to gravity, and θ is the angle of the inclined plane as in problem **16**.

71. In problem **25**; the magnitude of the component of the weight acting parallel to the inclined plane for the mass m_1 is $w_x = m_1\,g\,\sin\theta$.

72. In problem **25**; the magnitude of the the component of the weight acting parallel to the inclined plane for the mass m_2 is $w_x = m_2\, g\, sin\ \theta$.

73. In problem **34**; the component of the weight acting parallel to the inclined plane for the mass m_1 is $w_x = m_1\, g\, sin\ \theta_1$, downhill to the left.

74. In problem **34**; the component of the weight acting parallel to the inclined plane for the mass m_2 is $w_x = m_2\, g\, sin\ \theta_2$, downhill to the right.

75. In problem **38**; the component of the weight acting parallel to the inclined plane for the mass m_2 is $wx = m_2\, g\, sin\ \theta_2$.

76. In problem **46**; the component of the weight acting parallel to the inclined plane for the massless rope is zero.

77. In problem **58**; the component of the weight acting parallel to the inclined plane for the mass m_1 is $wx = m_1\, g\, sin\ \theta_1$, downhill to the right.

78. In problem **58**; the component of the weight acting parallel to the inclined plane for the mass m_2 is $wx = m_2\, g\, sin\ \theta_2$, downhill to the left.

79. For problem **16**, the normal force acts perpendicular to the inclined plane. This acts along **+y**.

80. For problem **16**, if the mass stays in contact with the inclined plane, the normal force N (do not confuse italic N, representing normal force, with roman N, which represents the newton a unit of force) of the inclined plane surface on the body has to balance the component of the weight perpendicular to the surface.

81. For problem **16**, the component of the weight perpendicular to the surface
$(= m\, g\, cos\ \theta)$ is along **–y**.

82. For problem **16**, $N = m\, g\, cos\ \theta$.

83. In problem **22**, $N + 5$ acts along the tilted **+y** and $m\, g\, cos\ \theta$ acts along **–y**.
$\Rightarrow N + 5 = m\, g\, cos\ \theta$.

84. For problem **25**, the normal force acting perpendicular to the surface on m_1 is
$N_1 = m_1\, g\, cos\ \theta$.

85.   For problem **25**, the normal force acting perpendicular to the
        surface on m_2 is
        $N_2 = m_2 g \cos \theta$.

86.   For problem **30**, the normal force acting perpendicular to the
        surface on m_1 is
        $N_1 = m_1 g \cos \theta$.

87.   For problem **30**, the normal force acting perpendicular to the
        surface on m_2 is
        $N_2 = m_2 g \cos \theta$.

88.   For problem **34**, the normal force acting perpendicular to the
        surface on m_1 is
        $N_1 = m_1 g \cos \theta_1$.

89.   For problem **34**, the normal force acting perpendicular to the
        surface on m_2 is
        $N_2 = m_2 g \cos \theta_2$.

90.   For problem **38**, the normal force acting perpendicular to the
        surface is on m_2
        $N_2 = m_2 g \cos \theta$.

91.   In the absence of friction, if an object slides down an inclined
        plane, the velocity and acceleration are in the same direction.
        The object speeds up.

92.   On a rough inclined surface, if a sliding object speeds up, then
        we can say that the acceleration and velocity are still in the
        same direction.

93.   If an object slows down while sliding down an inclined plane,
        then we can say that acceleration and velocity are in the
        opposite direction from each other.

Circular Motion

5

zaks!!

For problems **1—10**, a particle of mass m starts from point A and goes along a circular path of radius R with a uniform angular velocity ω. It goes from point A to point B in a time Δt.

1. The angular velocity has magnitude $\Delta\theta / \Delta t$. In what direction does it point?

   A. $\omega = (\Delta\theta) / (\Delta t)$. Perpendicularly out of the plane of the circle at point O.
   B. Along OA.
   C. Along OB.

2. What is the linear speed v_A at point A?

   A. $R \times \omega$ along AO.
   B. $R \times \omega$ along the tangent at point A.
   C. $R \times \omega$ along the perpendicular to the plane of the circle at point A.
   D. $R \times \omega$ along OA.

3. If $v_B - v_A = \Delta v$, the vector triangle representing these vectors is as shown in the figure ___. [Left to right —**A**, **B**, and **C**.]

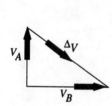 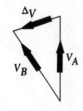 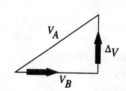

4. What is the direction of the change in velocity vector, Δv?

   A. Along OC.
   B. Toward O from midpoint of AB.
   C. Along AB.
   D. Along BA.

5. What is the relation connecting linear speed and angular speed?

   A. $v = R / \omega$.
   B. $v = \omega / R$.
   C. $v = R \times \omega$.

6. What is the acceleration of the particle? (It is all radial because ω is a constant. Tangential acceleration is zero.)

A. $R \times \omega^2$.
B. R / ω^2.
C. ω^2 / R.
D. $R^2 \times \omega$.

7. What is the direction of (radial) acceleration?

A. Toward the center along a radius.
B. Away from the center along a radius.
C. Along the tangent.
D. Perpendicular to the plane of the circle.

8. What is the change in speed along the tangent?

A. $2 \times v$.
B. Zero.
C. $v / 2$.

9. What is the tangential component of acceleration?

A. v^2 / R.
B. $v^2 / (2 \times AB_{\text{ARC LENGTH IN m}})$.
C. Zero.
D. $v^2 / AB_{\text{ARC LENGTH IN m}}$.

10. What is the net force acting on the particle?

A. $m \times R \times \omega^2$ toward the center.
B. $m \times R \times \omega^2$ away from the center.
C. $m \times g$ vertically down.
D. $m \times g$ vertically up.

For problems **11–13**, a particle moves around a circle with a uniform speed. Please respond with:

A. Yes, it is.
B. No, it is not.

11. Is angular velocity a constant?

12.   Linear momentum is defined as *mass × velocity*. Is linear momentum a constant?

13.   Is the acceleration toward the center a constant?

14.   When a car goes around a curve of a radius R with speed v, what is the coefficient of friction, μ (between the wheels of the car and the road), to prevent skidding?

   A. $\mu \geq v^2 / (R \times g)$.
   B. $\mu \leq v^2 / (R \times g)$.
   C. $\mu \geq v^2 / (2 \times R \times g)$.
   D. $\mu \leq v^2 / (2 \times R \times g)$.

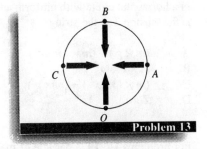

Problem 13

Physics Professors Don't Believe in Centrifugal forces.
by Joseph Straley

Whenever your car goes around a curve, you feel a force pulling you outwards. Your physics professor claims that it doesn't exist. So you go for a ride in your car — but just before the curve, your professor insists on getting out.

   It's not entirely due to the way you drive. Physics professors don't like being part of the experiment. Standing out on the road, they note that the direction of your motion is different before and after ("Your velocity vector has changed" is the official prof–speak for this event) and has changed by veering towards the center of the curve which every physics professor will insist implies a net force towards the inside of the curve.

   Meanwhile, if you stay in your car (recommended by the Department of Traffic Safety), you again experience a tendency to slide outwards — you want to say there is a force pulling you outwards. "No, no, no," cries your prof. "That is merely your tendency to continue moving in a straight line! No force involved." Fortunately, the professor is not in the car and is barely audible.

15. A body of mass m tied to a string of length l is whirled around in a horizontal circle with uniform angular velocity ω. What is the tension T in the string?

   A. $(m \times g) + (m \times l \times \omega^2)$.
   B. $m \times l \times \omega^2$.
   C. $(m \times l \times \omega^2) - (m \times g)$.
   D. Zero.

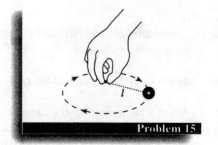
Problem 15

For problems **16–19**, a particle of mass m tied to a string of length l is whirled in a vertical circle with a uniform angular velocity ω.

16. What is the tension T in the string at point A?

   A. $(m \times g) + (m \times l \times \omega^2)$.
   B. $(m \times g) - (m \times l \times \omega^2)$.
   C. $(m \times l \times \omega^2) - (m \times g)$.
   D. Zero.

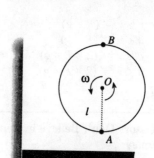

17. What is the tension T in the string at point B?

   A. $(m \times l \times \omega^2) + (m \times g)$.
   B. Zero.
   C. $(m \times l \times \omega^2) - (m \times g)$.
   D. $m \times g$.

18. What is the smallest possible value of ω such that the particle still reaches the top of the loop?

   A. Square root of $(g / l) = (g / l)^{1/2}$.
   B. Square root of $(l / g) = (l / g)^{1/2}$.
   C. Square root of $[(2 \times l) / g] = [(2 \times l) / g]^{1/2}$.

19. What is the condition for the particle to fall off the circle at point B?

   A. $\omega < (g / l)^{1/2}$.
   B. $\omega > (g / l)^{1/2}$.
   C. $\omega < (g \times l)^{1/2}$.
   D. $\omega > (g \times l)^{1/2}$.

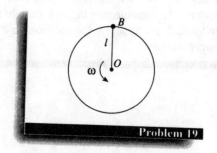
Problem 19

For problems **20-24**, a particle moves along a circle with a uniform speed *v* from A to B.

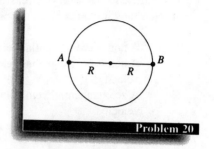

Problem 20

y

x

20. What is the time taken to go from A to B?

A. $(\pi \times R) / v$.
B. $(2 \times R) / v$.
C. R / v.
D. $(2 \times \pi \times R) / v$.

21. What is the change in tangential velocity as it goes from point A to point B? Assume **+x** is to the right, and **+y** vertically up.

A. Zero.
B. $-2 \times v$.
C. v.
D. $2^{1/2} \times v$.

22. What are the tangential accelerations at points A and B?

A. $v^2/(\pi \times R)$ and $v^2/(\pi \times R)$, respectively.
B. Zero and zero, respectively.
C. v^2/R and v^2/R, respectively.
D. Zero and $(2 \times v^2)/(\pi \times R)$, respectively.

23. What is the magnitude of the average radial acceleration? Assume a fixed coordinate system (**+x** to the right, **+y** vertically up) that is not rotating.

A. v^2/R.
B. Zero.
C. $(2/\pi) \times (v^2/R)$.
D. $(2 \times v^2)/R$.

24. (Assume a fixed coordinate system (**+x** to the right, **+y** vertically up) that is not rotating.) The radial acceleration at A and B are respectively:

A. $+v^2/R$ and $-v^2/R$.
B. Zero and zero.
C. $-v^2/R$ and $v^2/(\pi \times R)$.
D. Zero and $(2 \times v^2)/(\pi \times R)$.

25. If a particle moves along a circle of radius R with a variable speed, at any instant at any point on the circle, it has:

   A. A non—zero tangential velocity and radial velocity.
   B. A non—zero tangential velocity and zero radial velocity.
   C. A zero tangential velocity and non—zero radial velocity.
   D. A zero tangential velocity and zero radial velocity.

26. If a particle moves along a circle of radius R with a variable speed, at any instant at any point on the circle, it has:

   A. A zero tangential acceleration and zero radial acceleration.
   B. A non—zero acceleration and zero radial acceleration.
   C. A non—zero tangential acceleration and a non—zero radial acceleration.
   D. A zero tangential acceleration and a zero radial acceleration.

27. If a particle moves along a circle of radius R with a variable speed, what can we say about (the direction of) the resultant acceleration?

   A. It is along the tangent.
   B. It is radially inward.
   C. It is radially outward.
   D. It is at an oblique angle.

28. An open can filled with paint is revolved in a vertical circle by an attached string of length 0.4 m so the paint does not spill out. What should be the period of revolution to accomplish this safely?

   A. 1.26 s.
   B. 2.5 s.
   C. 0.63 s.

Problem 28

29. The acceleration due to gravity on the surface of the earth (radius 6,400 km) near the equator is ($g =$) 10 m/s^2. What is the centripetal acceleration of a mass at the equator due to the rotation of the earth?

   A. $3.4 \times 10^{-3} \times g$.
   B. g.
   C. $10^{-1} \times g$.

30. A car travels around an un—banked curve with a radius of 60 m without slipping at a speed of 54 km/hr. What is the least coefficient of friction μ, between the wheels and the road, to prevent skidding?

    A.   0.5.
    B.   0.375.
    C.   0.0375.
    D.   0.05.

31. The electron in the hydrogen atom orbits around the nucleus in a circular orbit of radius 0.528×10^{-10} m with a period of 15×10^{-15} s. What is the centripetal acceleration of the electron?

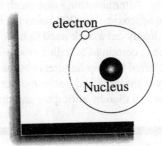

    A. 9×10^{18} m/s^2.
    B. 9×10^{-22} m/s^2.
    C. 9×10^{-16} m/s^2.
    D.   9 m/s^2.

32. A car of mass 1200 kg moves on a circular path with a constant speed of 54 km/hr from point A to point B (one quarter of a circle). If it travels for 314 m, what is the centripetal force on the car?

    A.   675 N.
    B.   1350 N.
    C.   2700 N.
    D.   350 N.

33. A circular disc is rotating about a vertical axis through its center with uniform angular velocity ω. A coin of mass m, is placed on the disc at a distance r from the center. What is the maximum value of r such that the coin will not be thrown off?

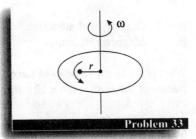

    A. $(\mu \times g) / \omega^2$.
    B. $(2 \times \mu \times g) / \omega^2$.
    C. $(\mu \times g) / (2 \times \omega^2)$.

Problem 33

Stored Energy
by Mike Slaughter

There are many safe ways to store energy, but every now and then a dangerous method must be used because it is unavoidable. Sailors working on the decks of large ships understand this and have a healthy respect for nylon. Lines (don't say *ropes* to a sailor, ropes are made of wire) made of nylon have a very high parting strength. A two-inch diameter line can hold up to 20,000 pounds of force before parting. They will also stretch about 30% before failure. After absorbing that much energy, a parted line will recoil at approximately 200 miles per hour, enough force (at contact) to cleanly cut a deck hand in half. Anyone in line with that whiplash is certain to be killed or seriously injured. Nylon lines are preferred by ship-handlers because they have a high strength-to-weight ratio, and you can observe the amount a nylon line stretches and thereby anticipate and prevent its failure.

34. If T is the period of revolution of a planet around the sun and R the radius of the circular orbit, then:

    A. $T \propto r$.
    B. $T^2 \propto r^3$.
    C. $T \propto r^2$.
    D. $T \propto r^3$.

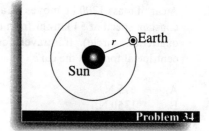

Problem 34

35. What provides the centripetal force when a planet orbits the sun?

    A.   Gravitational attraction.
    B.   Magnetic attraction.

36. What is the linear velocity of the earth's rotation at the equator (radius of the earth $= 6.4 \times 10^6$ m)?

    A.   473 m/s.
    B. 3×10^8 m/s.

37. What is the angular velocity of the earth's rotation at the equator?

    A. 73×10^{-6} rad/s.
    B. 146×10^{-6} rad/s.
    C. $2 \times \pi$ rad/s.

38. What is the angular velocity of the earth's rotation at the poles?

    A.   Zero.
    B. 73×10^{-6} rad/s.
    C. 146×10^{-6} rad/s.

39. What is the linear velocity of the earth's rotation at the poles?

    A.   473 m/s.
    B. 3×10^{8} m/s.
    C.   Zero.

40. A motorcyclist crosses a bridge which is an arc of a circle of radius 160 m. What is the safest speed of the motorcycle for it not to leave the road at the top of bridge?

    A.   Greater than 144 km / hr.
    B.   Less than 144 km / hr.
    C.   Less than 160 km / hr.
    D.   Greater than 160 km / hr.

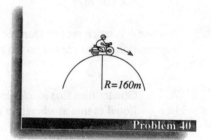

Problem 40

41. What is the force of gravitational attraction between the electron and the proton in a hydrogen atom? (Radius of orbit is 5×10^{-11} m, $m_e = 9 \times 10^{-31}$ kg, $m_p = 1.7 \times 10^{-27}$ kg, and $G = 6.7 \times 10^{-11}$.)

    A. 4.1×10^{-47} N.
    B.   10 N.
    C. 8×10^{-8} N.

42. It is estimated that the electron orbits 6.6×10^{15} times per second around the H nucleus. What centripetal force acts on the electron?

    A. 4.6×10^{-48} N.
    B. 8×10^{-8} N.
    C.   10 N.

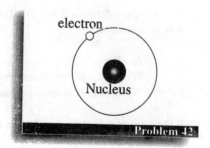

electron

Nucleus

Problem 42

43. What is the angular velocity of the earth around the sun (distance from the earth to the sun is 1.5×10^{11} m, mass of the sun is 1.99×10^{30} kg)?

    A. 2×10^{-7} rad/s.
    B. π rad/s.
    C. 2×10^{-6} rad/s.

44. What is the centripetal acceleration of the earth towards the sun?

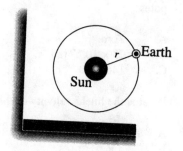

    A. $10 \, \mathrm{m/s}^2$.
    B. $6 \times 10^{-3} \, \mathrm{m/s}^2$.
    C. $6 \times 10^{5} \, \mathrm{m/s}^2$.

45. What is the gravitational acceleration of earth towards the sun?

    A. $10 \, \mathrm{m/s}^2$.
    B. $12 \times 10^{-5} \, \mathrm{m/s}^2$.
    C. $6 \times 10^{-3} \, \mathrm{m/s}^2$.
    D. $6 \times 10^{5} \, \mathrm{m/s}^2$.

46. Assume h is much greater than the radius of the earth. The acceleration due to gravity on the surface of the earth is:

    A.   Greater than that at a height h above the ground level.
    B.   Equal to that at a height h above the ground level.
    C.   Less than that at a height h above the ground level.

47. The acceleration due to gravity on the surface of the earth is:

    A.   Greater than that at a depth h below the earth's surface.
    B.   Equal to that at a depth h below the earth's surface.
    C.   Less than that at a depth h below the earth's surface.

48. When is the relation $v = R \times \omega$ valid? Assume v is in m/s and R is in meters.

    A.   At all times.
    B.   When ω is in revolutions per minute.
    C.   When ω is in radians per second.

49. If the ratio of mass and weight of a fixed object is $9.81 \, \mathrm{m/s}^2$ at the poles, then what is the value at the equator (considering the earth's rotation)?

    A. > 9.81.
    B. $= 9.81$.
    C. < 9.81.

50. A body is hung by a spring balance in a ship. What will the reading of the spring balance do as the ship moves from the poles to the equator?

    A. Increase.
    B. Decrease.
    C. Stay the same.

51. The direction of the rotational velocity of the earth at the equator is from:

    A. East to west.
    B. West to east.
    C. North to south.
    D. South to north.

52. For a geo-synchronous communication satellite, what is the period of revolution around the earth?

    A. One day.
    B. Less than one day.
    C. Greater than one day.

53. Which figure represents the variation of acceleration due to gravity with distance *r* from the center of the earth?

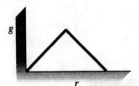 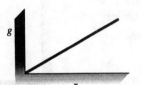  

| **Problem 53, Figure A** | **Problem 53, Figure B** | **Problem 53, Figure C** | **Problem 53, Figure D** |

54. The acceleration due to gravity on the moon is one sixth of that on the earth. If a sportsman can jump to a vertical height *h* on the earth, what is the height he can jump (with the same initial speed) on the moon?

    A. $6 \times h$.
    B. $h / 6$.
    C. $6^{1/2} \times h$.
    D. $h / 6^{1/2}$.

55. A man stands on a weighing scale in an elevator going down a mine with uniform speed. The reading of the scale will ____.

A. Increase.
B. Decrease.
C. Be steady.
D. First increase then decrease.

56. The variation of magnitude of the gravitational force F between two bodies with distance r is correctly represented by which graph?

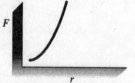

Problem 56, Figure A

Problem 56, Figure B

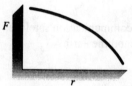

Problem 56, Figure C

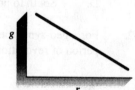

Problem 56, Figure D

For problems **57— 60**, an object of mass m tied to a thread is whirled in vertical circles. Choose one of the following equations of motion towards the center. Assume T is the tension in newtons, v is the speed in m/s, and R is the radius in meters.

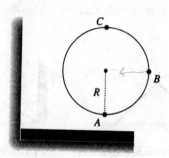

A. $T = m \times g.$
B. $T - mg = (mv^2) / R.$
C. $T + mg = (mv^2) / R.$
D. $T = (mv^2) / R.$
E. $mg = (mv^2) / R.$

57. What is the equation of motion at point A?

58. What is the equation of motion at point B?

59. What is the equation of motion at point C?

60. What is the equation of motion at point C if the circular velocity is so small that the mass just reaches point C? In the equation of motion, discard any force that is zero.

For problems **61—65**, a mass of *m* kg attached to a weightless cord of length *l* is suspended from a point O. It makes a cone as shown in the figure. The mass moves in a horizontal circle with constant angular speed ω. The string keeps a constant angle α with the vertical (θ with the horizontal).

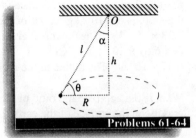

Problems 61-64

61.   What is the radius of the circle, *R*?

    A.   *l*.
    B.   *l* × *sin α*.
    C.   *l* × *cos α*.
    D.   *l* × *tan α*.

62.   What is the tension of the string?

    A.   *m* × *g*.
    B.   *m* × *g* × *sin θ*.
    C.   (*m* × *g*) ⁄ *cos α*.
    D.   (*m* × *g*) ⁄ *sin α*.

63.   What happens as the angular speed (= ω) of the mass increases?

    A.   It will remain in the same position.
    B.   It will be raised (*h* decreases).
    C.   It will be lowered (*h* increases).

64.   What happens to the tension *T* as we raise the object?

    A.   It stays the same.
    B.   It will increase.
    C.   It will decrease.

65.   What is the period of revolution of the mass?

    A. $2 \times \pi \times [(l \times sin\ \alpha) / g)]^{1/2}$.
    B. $2 \times \pi \times [(l / g)]^{1/2}$.
    C. $2 \times \pi \times [(l \times cos\ \alpha) / g)]^{1/2}$.
    D. $2 \times \pi \times [(g / l)]^{1/2}$.

The following information should help you to do problems
66–68. The earth-moon distance is 3.8×10^8 m. The mass of
the earth is 6×10^{24} kg. Mass of the moon is
7.38×10^{22} kg. The universal gravitational constant is
$G = 6.67 \times 10^{-11}$ in MKS units. The radius of the earth is
6.37×10^6 m. The radius of the moon is 1.74×10^6 m.

66.   What is the approximate ratio of the acceleration due to
         gravity on the surface of the earth to that on the moon?

         A.   5.
         B.   1 / 8.
         C.   10.
         D.   1 / 36.

67.   What is the ratio of centripetal acceleration to the gravitational
         acceleration on the surface of the earth (approximately)?

         A.   1.
         B. 3.4×10^{-3}.
         C.   10.

You
Can Do
It!

68.   What is the ratio of the centripetal force on the moon around
         the earth to the gravitational force between the moon and
         earth?

         A.   1.00.
         B.   0.5.
         C.   1.5.
         D.   2.

69.   Two earth satellites are orbiting the earth at heights h_1 and h_2
         above the surface of the earth. What is the ratio of their
         periods of revolution T_1 / T_2?   Assume r_e is the radius of
         the earth.

         A. $[(r_e + h_1) / (r_e + h_2)]^{3/2}$.
         B. $[(r_e + h_1) / (r_e + h_2)]^{1/2}$.
         C. $(r_e + h_1) / (r_e + h_2)$.
         D. $(r_e + h_2) / (r_e + h_1)$.

70. What is the ratio of acceleration due to gravity on the surface of the earth g_e to that at a height h above the surface of the earth g_h?

A. $(r_e + h) / r_e$.
B. $(r_e + h)^2 / r_e^2$.
C. $r_e^2 / (r_e + h)^2$.
D. $r_e / (r_e + h)$.

71. For a satellite orbiting the earth at a small height (i.e. an earth satellite whose height, h, is much less than the radius of the earth, R_E. $\Rightarrow$ Gravity does not change much), what is the orbital velocity?

A. $(2 \times g_E \times R_E)^{1/2}$.
B. $(g_E \times R_E)^{1/2}$.
C. $[(g_E \times R_E) / 2]^{1/2}$.

72. For problem **71**, what is the period of revolution of the satellite?

A. $2 \times \pi \times (R_E / g)^{1/2}$.
B. $2 \times \pi \times (g / R_E)^{1/2}$.
C. $2 \times \pi \times (2 \times R_E / g)^{1/2}$.

73. An object can move in a straight line if the net force on it acts in the direction of motion.

74. An object can move in a straight line if the net force on it is zero.

75. An object can move in a straight line if the vector sum of multiple force components adds to zero.

76. An object can move in a curved path if the net force on the object is not zero.

77. If the net force on an object is zero, the object can never move in a curved path.

78. If an airplane moves in a circular path before it lands, then net force on it is not zero.

79. If the earth moves in a circular path around the sun, then the net force on the earth is not zero.

80. If a space shuttle moves in a circular path around the earth, then net force on it is not zero.

81. If a car moves in a straight line at constant velocity, the net force on the car is zero.

82. Uniform circular motion is when an object moves in a circle at constant speed (that is, magnitude of the velocity vector as it goes around in circles stays the same but not its direction).

83. If an object is undergoing uniform circular motion, it is continuously accelerating.

84. The magnitude of the acceleration of an object undergoing uniform circular motion = v^2 / R where v is the speed in m/s and R is the radius in m.

85. The magnitude of the net force of an object undergoing uniform circular motion = $(mv^2)/R$ where m is the mass in m/s, v is the speed in m/s, and R is the radius in m.

86. The acceleration vector of an object undergoing uniform circular motion is directed towards the center of the circle.

87. The acceleration towards the center of the circle of an object undergoing uniform circular motion is called centripetal acceleration or radial acceleration.

88. An object that moves in a curved path will have radial or centripetal acceleration.

89. The velocity vector always points in the direction of motion.

90. Velocity and acceleration vectors are perpendicular to each other at every point during uniform circular motion.

91. The average speed of an object that moves in a circular path = circumference / period, where circumference is the distance traveled in m for one revolution and period is the time it takes to complete one revolution.

92. The average speed of the earth around the sun (assume a circular orbit for convenience) = circumference / period, where circumference = $2\pi R$ is the distance traveled by the earth around the sun for one revolution and T is the period of the earth around the sun (= 1 year).

93. If a needle takes one second to complete one revolution, then its period is 1 s.

94. If the earth takes 24 hours to complete one rotation about its axis, we say that the period of the earth about its own axis is 24 hours.

95. If the earth takes a year to complete one revolution around the sun, then we say that the period of the earth around the sun is 1 year.

96. Centripetal acceleration for an object undergoing uniform circular motion depends upon the speed and radius.

97. If the centripetal acceleration for an object undergoing uniform circular motion is v^2/R, then the bigger the speed, the faster the velocity vector changes the direction provided the radius R is a constant.

98. If centripetal acceleration for an object undergoing uniform circular motion is v^2 / R, then the bigger the radius, the slower the velocity vector changes its direction provided v is a constant.

99. The net force towards the center of the circle is $(mv^2)/R$ for an object undergoing (uniform) circular motion.

100. The net force towards the center of a circle for an object undergoing uniform circular motion is the vector sum of the force components that act along the line joining the center and the object's position at a distance equal to the radius of the circle R in m.

101. Centripetal acceleration can occur due to just one force acting on an object that is moving in a circle.

102. Centripetal acceleration can occur due to multiple forces acting on an object that is moving in a circle.

103. Your incoming speed at a highway's curved exit ramp is not a constant if you slow down from 65 mph to 15 mph. You then have tangential acceleration in addition to centripetal acceleration.

104. If an object moves in a circle at constant speed, its tangential acceleration is zero.

105. If an object moves in a circle at constant speed, its centripetal acceleration is not zero.

106. In problem **105**, we have both centripetal and tangential acceleration.

107. If the earth's orbit is a circle around the sun, then earth's tangential acceleration is zero but not its centripetal acceleration if speed is a constant.

108. For problem **15**, tension T acts horizontally to the left towards the center of the circle. $\Rightarrow T = (mv^2) / R \Rightarrow T = (mv^2) / l$.

109. For problem **16**, at point A at the bottom, the tension acts towards the center and the weight acts away from the center. The net force is then:
$$T - (mg) = (mv^2) / R. \Rightarrow T - (mg) = (mv^2) / l.$$

110. For problem **16**, at point B at the top, the tension acts towards the center and the weight also acts towards the center. The net force towards the center is then:
$T + (mg) = (mv^2) / R$. $\Rightarrow T + (mg) = (mv^2) / l$.

111. In problem **40**, the normal force exerted by the surface on the motorcycle is vertically upwards (radially out). Weight acts vertically down towards the center of the circle. The equation of motion if the motorcycle is in contact with the surface becomes:
$(mg) - N = (mv^2) / R$.

112. For problem **61**, the horizontal component of the tension is to the right towards the center of the circle. The net force towards the center of the circle is then:
$T \cos \theta = (mv^2) / R$.
$\Rightarrow T \cos \theta = (mv^2) / (l \cos \theta)$.

113. For problem **61**, the vertical component of the tension $T \sin \theta$ is vertically up.

114. For problem **61**, the weight mg is vertically down.

115. For problem **61**, if the net force along the vertical is zero, then $T \sin \theta = mg$.

116. The sun pulls on the earth just as the earth pulls on the sun.

117. The force between two masses is always attractive if we ignore contact forces.

118. When the earth revolves around the sun, the force towards the center is the gravitational force of attraction between the earth and the sun.

119. Gravitational forces between two point masses reduce with increase in distance
as $1 / r^2$.

Work & Energy

6

For problems **1-14**, choose one of the following as the correct answer. (Neglect friction and air resistance unless otherwise stated).

A. Total energy at the start is all potential energy. As it is completely converted to kinetic energy, there will be a gain in speed.

m MASS g ACCE. GRAVITY h HEIGHT FROM THE ZERO LEVEL = (½) m MASS v VELOCITY ^{2}.

B. Total energy is all kinetic energy at the start. As this kinetic energy gets converted to all potential energy, there will be a decrease in speed.

(½) m MASS v VELOCITY ^{2} =
m MASS g ACCE. GRAVITY h HEIGHT FROM THE ZERO LEVEL.

C. At the start we have both kinetic energy and potential energy. As potential energy decreases there will be a further gain in speed.

(½) m MASS v INITIAL VELOCITY ^{2} +
m MASS g GRAVITY h HEIGHT FROM THE ZERO LEVEL =
(½) m MASS v FINAL VELOCITY ^{2} if the final height from the zero level of potential energy is zero.

1. A child (system) falls off the bed at night from rest. Ouch!
[Start = at rest on the bed. Finish = just before hitting the ground.]

2. A person (system) steps on the ice and falls flat hitting her head in the process. Assume she has a non-zero speed before falling.
[Start = walking at a non-zero speed. Finish = just before hitting the ground.]

3. A ball (system) is kicked horizontally from the roof of a building.
[Start = just after it is kicked. Finish = just before hitting the ground.]

4. A ball (system) thrown vertically up climbs to the maximum height.
[Start = just after the release. Finish = at the instant the ball reaches the maximum height.]

5. A ball (system) is dropped from rest.
[Start = at rest. Finish = just before hitting the ground.]

6.   A skier (system) comes down to the base of a hill. Assume he was moving initially with a non-zero speed.
[Start = at the top of the hill. Finish = just before reaching the base.]

7.   A car (system) initially at rest slides down an inclined plane to the base of the hill.
[Start = at the top of the hill. Finish = just before reaching the base.]

8.   A motorcyclist (system) climbs up a ramp after she turns off the engine at the base, until she stops at a higher elevation point.
[Start = at the base of the hill. Finish = at the instant she reaches the top.]

9.   A roller coaster (system) comes down from the top of a vertical loop.
[Start = at the top. Finish = just before reaching the bottom.]

10.   A person (system) on a swing is on the way up.
[Start = at the bottom. Finish = at the instant it reaches the maximum height.]

11.   A mass (system) on a pendulum released at rest from a height h is on the way down.
[Start = at maximum height. Finish = at the instant the person reaches the bottom point.]

12.   A golf ball (system) rolls around the rim until it falls into the hole (cup).
[Start = at the top as it rolls. Finish = just before hitting the bottom of the cup.]

13.   A bullet (system) is fired horizontally.
[Start = at the exit from the gun. Finish = just before it hits the ground.]

14.   The City Building crumbles after it was bombed.
[Start = at rest before it comes down. Finish = at the instant it reaches the ground.]

Define work done by any force as:

$|force|_{\text{MAGNITUDE}} \times |displacement|_{\text{MAGNITUDE}} \times cos\ \theta_{\text{ANGLE}}$
BETWEEN THE DIRECTION OF DISPLACEMENT AND DIRECTION OF THAT
PARTICULAR FORCE·

For problems **15-30**, choose one of the following answers.

A.   Work done by the gravitational field on the system is zero.

[For example: Work done by the gravitational field = $|m_{\text{MASS}}\ g_{\text{ACCE. GRAVITY}}\ y_{\text{VERTICAL DISPLACEMENT}}| \times cos\ 90$. Also if the vertical displacement is zero, then the work done by a force can also be zero.]

B.   Work done by the gravitational field on the system is positive (a loss of potential energy; the field does the work for you).

[For example: Work done by the gravitational field = $|m_{\text{MASS}}\ g_{\text{ACCE. GRAVITY}}\ y_{\text{VERTICAL DISPLACEMENT}}| \times cos\ 0$.]

C.   Work done by the gravitational field on the system is negative (potential energy increases; you give energy to the field, or we can say we do work against the field).

[For example: Work done by the gravitational field = $|m_{\text{MASS}}\ g_{\text{ACCE. GRAVITY}}\ y_{\text{VERTICAL DISPLACEMENT}}| \times cos\ 180$.]

15.   An ice skater (system) slides to rest on the ice. Assume the surface is horizontal.

16.   As you open the shower pipe, water (system) spurts out and down.

17.   You spill coffee (system) on the table.

18.   You take a 32-inch TV (system) upstairs.

19.   Water (system) drains off a storage tank.

20.   A person (system) does a high jump (she is on the way up).

21.   A trapeze performer (system) falls to the net.

22.   A hen is laying an egg (system) onto the ground.

23.   A bungee jumper (system) leaps off a bridge.

24.   You flush the water (system) in the toilet on the 120th floor.

25.   You put change (system) into your pocket.

26.   You open the tap to wash your hands with water (system) and soap.

27.   We melt the butter (system) to make corn pudding.

28.   After unpacking, a person throws the boxes (system) down out the window from the third floor.

29.   A space shuttle (system) moves in a circular path of constant radius around the earth.   (Assume for simplicity that the earth is a sphere.)

30.   An airplane (system) takes off.

For problems **31–34**, a person carries a suitcase of mass 15 kg at a height of 0.2 m in her hand while she waits at a bus stop to catch a bus. The bus stops 50 m away from her.

31.   What is the work done on the suitcase (system) when she is waiting for the bus?

A.   Zero.
B.   30 J.
C.   3 J.
D.   150 J.

32.   What is the work done on the suitcase (system) by the force of gravity if she walks with a uniform speed of 6 km/hr to board the bus?

A.   750 J.
B.   7500 J.
C.   Zero.
D.   15 J.

33.   What is the work done by the force of gravity on the suitcase (system) when she boards the bus with its floor at a height of 0.5 m above the ground?

A.   45 J.
B.   Zero.
C.   75 J.
D.   7.5 J.

34. What is the work done on the suitcase (system) by all the forces acing on the system (called the net work done by the net force) if she starts running with uniform acceleration and reaches the bus in 15 s?

    A.   333 J.
    B.   3330 J.
    C.   1500 J.
    D.   Zero.

For problems **35-57**, consider the projectile shown in the figure. Find the kinetic energy, potential energy, and the total energy at the points indicated. Ignore air resistance.

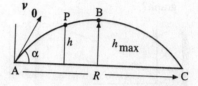

35. For point A:
[KE KINETIC ENERGY, PE POTENTIAL ENERGY, TE TOTAL ENERGY]

    A.   Zero, $(\frac{1}{2} \times m \times v_0^2)$, $(\frac{1}{2} \times m \times v_0^2)$.
    B. $(\frac{1}{2} \times m \times v_0^2)$, zero, $(\frac{1}{2} \times m \times v_0^2)$.
    C. $(\frac{1}{2} \times m \times v_0^2)$, $(-m \times g \times h)$, Zero.
    D. $(-m \times g \times h)$, zero, $(\frac{1}{2} \times m \times v_0^2)$.

Problems 35-57

36. For point P:
[KE KINETIC ENERGY, PE POTENTIAL ENERGY, TE TOTAL ENERGY]

    A. $[(\frac{1}{2} \times m \times v_0^2) - (m \times g \times h)]$, $(m \times g \times h)$,
        $(\frac{1}{2} \times m \times v_0^2)$.
    B. $(\frac{1}{2} \times m \times v_0^2 \times cos^2 \alpha)$, $(m \times g \times h \times sin\, \alpha)$,
        $(m \times g \times h)$.
    C. $(\frac{1}{2} \times m \times v_0^2)$, $(m \times g \times h)$, $(\frac{1}{2} \times m \times v_0^2)$.
    D.   Zero, $(m \times g \times h)$, $(m \times g \times h)$.

37. For point B (assume h_m is the maximum height):
(KE KINETIC ENERGY, PE POTENTIAL ENERGY, TE TOTAL ENERGY).

    A. $(\frac{1}{2} \times m \times v_0^2)$, $(m \times g \times h_m)$, $(\frac{1}{2} \times m \times v_0^2)$.
    B. $(\frac{1}{2} \times m \times v_0^2 \times sin^2 \alpha)$, $(m \times g \times h_m)$,
        $(\frac{1}{2} \times m \times v_0^2 \times cos^2 \alpha)$.
    C. $(\frac{1}{2} \times m \times v_0^2 \times cos^2 \alpha)$, $(m \times g \times h_m)$, $(\frac{1}{2} \times m \times v_0^2)$.
    D. $(\frac{1}{2} \times m \times v_0^2 \times sin^2 \alpha)$, zero, $(m \times g \times h_m)$.

38. For point C (just before the projectile strikes the ground):
(KE KINETIC ENERGY, PE POTENTIAL ENERGY, TE TOTAL ENERGY)]

    A.   Zero, zero, zero.
    B.   ($\frac{1}{2} \times m \times v_0^2 \times cos^2\,\alpha$), zero, ($\frac{1}{2} \times m \times v_0^2 \times sin^2\,\alpha$).
    C.   ($\frac{1}{2} \times m \times v_0^2 \times sin^2\,\alpha$), ($m \times g \times h_m$), zero.
    D.   ($\frac{1}{2} \times m \times v_0^2$), zero, ($\frac{1}{2} \times m \times v_0^2$).

39. For a projectile, the variation of potential energy V with the horizontal displacement x is represented correctly by which graph?

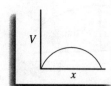 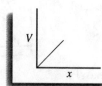 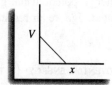 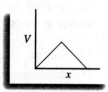

   Problem 39 - A   **Problem 39 - B**   **Problem 39 - C**   **Problem 39 - D**

40. For the projectile, the variation of kinetic energy T with horizontal displacement x is correctly represented by which graph?

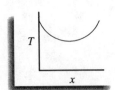 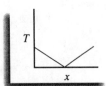 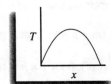 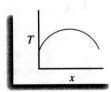

   Problem 40 - A   **Problem 40 - B**   **Problem 40 - C**   **Problem 40 - D**

41. For the projectile, the variation of total energy E with horizontal distance x is correctly represented by which graph?

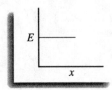 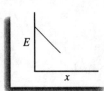 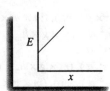 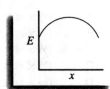

   Problem 41 - A   **Problem 41 - B**   **Problem 41 - C**   **Problem 41 - D**

42. If the curve ABC in the figure represents the variation of kinetic energy T with the horizontal displacement x of a projectile for angle of projection α, then:

    A. $\tan^2 \alpha = (BM / AO)$.
    B. $\cos^2 \alpha = (BM / AO)$.
    C. $\sin^2 \alpha = (BM / AO)$.
    D. $\cot^2 \alpha = (BM / AO)$.

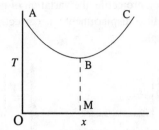

Problem 42

43. For a projectile with horizontal range R, the potential energy is maximum at:

    A.   The point of projection.
    B.   A distance $R / 2$.
    C.   A distance R.
    D.   A distance $(3 \times R)/ 2$.

44. For a projectile with a horizontal range R, the kinetic energy is maximum at:

    A.   The point of projection.
    B.   A distance $R / 4$.
    C.   A distance $R / 2$.
    D.   A distance $(3 \times R)/ 2$.

45. For a projectile with a horizontal range R, the total energy is equal to the kinetic energy at:

    A.   The point of projection.
    B.   A distance $R / 4$.
    C.   A distance $R / 2$.
    D.   A distance $(3 \times R)/ 2$.

46. For a projectile kicked at an angle to the horizontal, the total energy can be purely potential at some point on the path.

    A.   Yes.
    B.   No.

47. For a projectile, the variation of potential energy V with vertical displacement y is correctly represented by which graph?

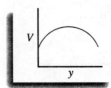 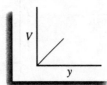 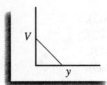 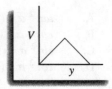

| **Problem 47 - A** | **Problem 47 - B** | **Problem 47 - C** | **Problem 47 - D** |

48. For a projectile, the variation of kinetic energy T with vertical displacement y is correctly represented by which graph?

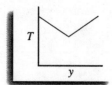 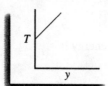 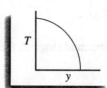 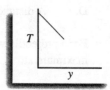

| **Problem 48 - A** | **Problem 48 - B** | **Problem 48 - C** | **Problem 48 - D** |

49. For a projectile, the variation of total energy E with vertical displacement y is correctly represented by which graph?

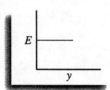 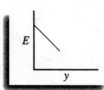 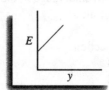 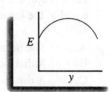

| **Problem 49 - A** | **Problem 49 - B** | **Problem 49 - C** | **Problem 49 - D** |

50. If the variation of kinetic energy of a projectile with vertical displacement is given by the graph shown in figure, then the angle α is determined by:

A. $tan^2 \alpha = (BM / AO)$.
B. $cos^2 \alpha = (BM / AO)$.
C. $sin^2 \alpha = (BM / AO)$.

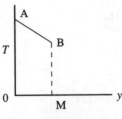

Problem 50

51. For a projectile, the minimum value of *KE* is:

   A. $\frac{1}{2} \times m \times v_0^2$.
   B.   Zero.
   C. $\frac{1}{2} \times m \times v_0^2 \times \cos^2 \alpha$.
   D. $\frac{1}{2} m \times v_0^2$.

52. For a projectile, the maximum value of *PE* is: [Hint: Use the trig identity $\sin^2 \alpha + \cos^2 \alpha = 1$]

   A. $\frac{1}{2} \times m \times v_0^2$.
   B. $\frac{1}{2} \times m \times v_0^2 \times \sin^2 \alpha$.
   C. $\frac{1}{2} \times m \times v_0^2 \times \cos^2 \alpha$.
   D. $m \times v_0^2$.

53. For a projectile, the variation of *V* (*PE*) with time *t* is correctly represented by which graph?

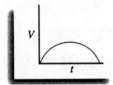   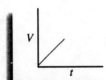   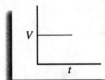   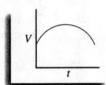

Problem 53 - A   **Problem 53 - B**   **Problem 53 - C**   **Problem 53 - D**

54. For a projectile, the variation of kinetic energy *T* with time *t* is correctly represented by which graph?

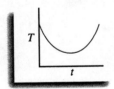   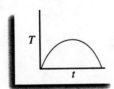   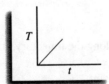   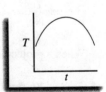

Problem 54 - A   **Problem 54 - B**   **Problem 54 - C**   **Problem 54 - D**

55.   For a projectile, the variation of total energy E with time t is given by which graph?

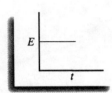

Problem 55 - A   **Problem 55 - B**   **Problem 55 - C**   **Problem 55 - D**

For problems **56-57**, for a projectile, the variation of kinetic energy T with time t is given by the accompanying graph.

56.   What does OA represent?

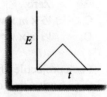

        A.   The initial energy.
        B.   The total PE.
        C.   The minimum PE.

57.   What does AC represent?   **Problems 56-57**

        A.   The total energy.
        B.   The PE at maximum height.
        C.   The KE at maximum height.
        D.   The difference between kinetic and potential energy.

58.   As a general statement, is potential energy always positive?

        A.   Yes.
        B.   No.

59.   When a projectile goes along the trajectory, its:

        A. PE remains constant.
        B. KE remains constant.
        C.   (TE) Total energy remains a constant.
        D.   Horizontal component of kinetic energy remains
                constant.

60. Two bodies (1) and (2) are released from rest at C, one let go on a smooth inclined plane and the other dropped from point C to point B (vertically). Ignore friction. One slides along CA and the other falls along CB. If the speeds they attain are respectively v_1 and v_2 at the end of their flight, then:

A. $v_1 > v_2$.
B. $v_1 < v_2$.
C. $v_1 = v_2$.

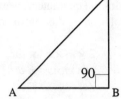

Problem 60

61. Use the figure of problem **60**, and assume no friction. If w_1 is the work required to lift a body to the point C of an inclined plane along AC, and w_2 is that for lifting it along BC, then:

A. $w_1 = w_2$.
B. $w_1 > w_2$.
C. $w_1 < w_2$.

62. The area under the curve connecting force and displacement gives:

A.   The average velocity.
B.   The work done.
C.   The power.
D.   The impulse.

63. When a body (the system) goes up a smooth inclined plane, work is done on the system:

A.   Against gravity.
B.   Against the normal reaction of the plane.
C.   By the normal reaction of the plane.

64. When a body (the system) moves down a smooth inclined plane, work is done on the system:

A.   By gravity.
B.   By the normal reaction of the plane.
C.   Against gravity.
D.   Against the normal reaction of the plane.

65. When a particle (the system) of mass m goes around a circle of radius r with uniform angular speed ω (it sweeps equal angles in equal intervals of time), the work done during half a revolution by the centripetal force is:

A. $2 \times m \times r^2 \times \omega^2$.
B. $\pi \times m \times r^2 \times \omega^2$.
C. Zero.

66. When the earth revolves in an approximately circular orbit around the sun with speed v, the work done by the sun on the earth is:

A. Zero.
B. $\frac{1}{2} \times m_{EARTH} \times v^2$.
C. $\frac{1}{2} \times m_{SUN} \times v^2$.
D. $[(m_{EARTH} + m_{SUN}) / 2] \times v^2$.

For problems that involve horsepower, remember that one horsepower – 1 hp = 750 Watts/sec.

67. A truck of mass m kg with power P moves along a horizontal road with a coefficient of friction μ at uniform speed. What is the maximum speed it can go in m/s if the road is level?

A. $(\mu \times m \times g) / P$.
B. $P / (\mu \times m \times g)$.
C. $(P \times \mu) / (m \times g)$.
D. $(m \times g) / (\mu \times P)$.

68. For the above problem, assume the road is an inclined plane and that the truck is climbing up. Assume α is the angle the inclined plane makes with the horizontal.

A. $P / [(m \times g) \times (\mu \times cos\, \alpha + sin\, \alpha)]$.
B. $P / [(\mu \times m \times g \times sin\, \alpha)]$.
C. $P / [(m \times g) \times (\mu \times sin\, \alpha + cos\, \alpha)]$.
D. $[(m \times g) \times (\mu \times sin\, \alpha + cos\, \alpha)] / P$.

69. If v_1 is the maximum speed a car with coefficient of friction μ can develop on a level road, and v_2 is it's maximum speed up along an incline of angle α to the horizontal (assume the same power), then v_1 / v_2 is:

    A. $(\mu \times cos\ \alpha - sin\ \alpha) / \mu$.
    B. $(\mu \times cos\ \alpha + sin\ \alpha) / \mu$.
    C. $(\mu \times sin\ \alpha + cos\ \alpha) / \mu$.
    D. $(\mu \times sin\ \alpha - cos\ \alpha) / \mu$.

70. Can kinetic energy ever be negative?

    A. Yes.
    B. No.

71. Can total energy ever be negative?

    A. Yes.
    B. No.

72. Unit of power can be in measured in joules per sec.

    A. True.
    B. False.

73. The relation between the momentum p ($= mv$) and kinetic energy T of a body of mass m is:

    A. $T = p^2 / m$.
    B. $T = p^2 / (2 \times m)$.
    C. $T = (2 \times m) / p^2$.
    D. $T = (2 \times m) / p^2$.

74. The kinetic energy of a body can be:

    A. Positive or zero.
    B. Negative.
    C. Positive, zero, or negative.

75. The potential energy of a body can be:

    A. Positive.
    B. Negative.
    C. Positive, zero, or negative.

76. The total energy of a body can be:

A. Positive.
B. Negative.
C. Positive, zero, or negative.

77. For a body with potential energy V, kinetic energy T, and total energy E, the condition for motion (to attain a non-zero speed) is:

A. $E > V$.
B. $E < V$.
C. $E = V$.
D. $E < T$.

78. A body with total energy E and potential energy V will be at rest when:

A. $E > V$.
B. $E < V$.
C. $E = V$.

79. A small mass m hung vertically by a massless string (of length l) from a point A is displaced from B to C so that the string makes an angle θ with the vertical. The change in potential energy PE of the mass is:

A. $m \times g \times l \times (1 - cos\ \theta)$.
B. $m \times g \times l \times (1 - sin\ \theta)$.
C. $m \times g \times l \times (cos\ \theta + 1)$.
D. $m \times g \times l \times (sin\ \theta + 1)$.

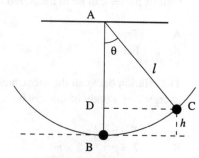

Problem 79

80. A uniform heavy stick of length L is hung vertically from a point A. When displaced through an angle θ from the vertical, its change of potential energy is:

A. $m \times g \times (L / 2) \times (1 + cos\ \theta)$.
B. $m \times g \times (L / 2) \times (1 - cos\ \theta)$.
C. $m \times g \times (L / 2) \times (1 + sin\ \theta)$.
D. $m \times g \times (L / 2) \times (1 - sin\ \theta)$.

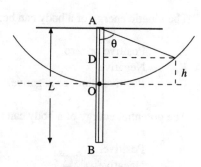

Problem 80

81. If F is the force required to raise a mass vertically through a height h and F_1 is that required to raise the same mass through the same vertical height but along an inclined plane of angle α, then (neglecting friction):

    A. $F < F_1$.
    B. $F > F_1$.
    C. $F_1 = F$.

82. A box of mass 40 kg is dragged horizontally along a level frictionless floor through 5 m and then raised 2 m. What is the magnitude of the work done by gravity?

    A.   2800 J.
    B.   800 J.
    C.   2000 J.
    D.   1200 J.

83. A box of mass 40 kg is dragged horizontally along a level floor with friction coefficient 0.3 through 5 m and then raised 2 m. What is magnitude of the net work done?

    A.   1400 J.
    B.   2800 J.
    C.   800 J.
    D.   600 J.

84. A steel ball of mass 1 kg falls from rest to the ground through a vertical height of 40 m and penetrates 0.2 m in the ground. The average resistance of the ground is:

    A.   400 N.
    B.   800 N.
    C.   2000 N.
    D.   4000 N.

85. A heavy uniform rope of mass m and length l hanging vertically from the roof of a building is hauled to the roof. The work done is:

    A. $m \times g \times l$.
    B. $m \times g \times (l / 2)$.
    C. $m \times g \times (l / 4)$.
    D. $(3 \times m \times g \times l) / 4$.

86. For a body thrown vertically upwards, its direction of motion changes at the point where its total energy is:

    A. Greater than the *PE*.
    B. Less than the *PE*.
    C. Equal to the *PE*.

87. When the force is conservative, the work done by the force (for a complete closed cycle):

    A. Depends on the path of motion.
    B. Does not depend on the path of motion.
    C. Is negative.
    D. Is positive.

88. An example of conservative force is:

    A. Gravitational force.
    B. Frictional force.
    C. Viscous force.

89. A body is raised from point A on the earth to point B above the earth's surface. The work done to lift the mass along ACB is:

    A. More than that along AB.
    B. Less than that along ADB.
    C. More than that along AEB.
    D. The same for all the paths.

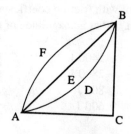

Problem 89

For problems **90-91**, a body falling from rest at a height of h rebounds to a height of $(3 \times h) / 4$ after touching the ground.

90. The percentage energy lost during the collision with the ground is:

    A. 42.
    B. 25.
    C. 75.
    D. Zero.

91. The velocity with which it rebounds is:

    A. Square root of $(2 \times g \times h)$.
    B. Square root of $(3 \times g \times h) / 4$.
    C. Square root of $[3 \times g \times (h / 2)]$.
    D. $2 \times$ square root of $(g \times h)$.

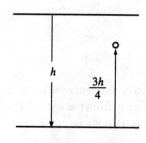

Problems 90-91

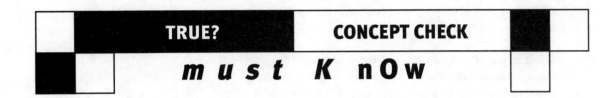

92. Work and energy are scalar quantities and therefore have no direction associated with them.

93. Work is accomplished by the application of a force when the object is moved through a distance.

94. If you push on the wall, no movement occurs. In the language of physics, we say that the application of your force did no work (you may say otherwise).

95. If you carry a book (levelly) on your head, no work is produced by the gravitational force because there is no displacement in the direction of the force.

96. A constant force means both magnitude and direction are constant.

97. Work done on an object by a constant force can be described as $w = F \times d_{\parallel}$. Work is force F times displacement $d_{\parallel}$ in the direction of the force.

98. The gravitational force acting on an object very close to the surface of the earth is an example of a constant force.

99. In general, work done by a force is $w = F\, d\, cos\, \theta$, where F is the magnitude of the force, d is the magnitude of the displacement, and θ is the angle between the direction of force and the direction of displacement.

100. The angle θ between the direction of force and direction of displacement is zero when force and displacement are in the same direction.

101. When an object falls vertically down towards the earth, the object's displacement and the gravitational force are both in the same direction. The angle θ between the direction of force and the direction of displacement is zero.

102. The angle θ between the direction of force and the direction of displacement is 180 degrees when force and displacement are in opposite directions.

103. When an object is thrown upward, the gravitational force acts vertically down but vertical displacement initially is in the opposite direction. This makes the angle θ between the direction of force and the direction of displacement = 180 degrees.

104. When a grocery cart initially at rest is pushed with a certain constant force on a horizontal surface, and it moves through a certain distance, both displacement and the direction of the force applied are in the same direction. The angle θ between the direction of the force applied and the direction of displacement is zero.

105. Work done by a force $w = F\, d \cos \theta$ (where F is the magnitude of the force, d is the magnitude of the displacement, and θ is the angle between the direction of force and the direction of displacement) is positive when $\cos \theta$ is positive because F and d are magnitudes and thus always positive or zero.

106. Work done by a force $w = F\, d \cos \theta$ is negative when $\cos \theta$ is negative.

107. If $\theta = 0$ degree, $\cos 0 = 1$. $\Rightarrow w = F\, d \cos \theta$ is positive when force and displacement are in the same direction.

108. If $\theta = 180$ degrees, $\cos 180 = -1$. $\Rightarrow w = F\, d \cos \theta$ is negative when force and displacement are in the opposite directions.

109. If $\theta = 90$ degrees, $\cos 90 = 0$. $\Rightarrow w = F\, d \cos \theta$ is zero when force and displacement are at right angles.

110. Along an inclined plane, the magnitude of the component of the weight parallel to the inclined plane is $mg \sin \alpha$, where m is the mass in kg, g is the magnitude of the acceleration due to gravity, and α is the angle of the inclined plane with respect to the horizontal.

111. If an object slides down a frictionless inclined plane, the "downward" force $mg \sin \alpha$ parallel to the inclined plane and displacement are in the same direction.

112. If the force $mg \sin \alpha$ parallel to the inclined plane and displacement are in the same direction, $\theta = 0$. $\Rightarrow$ $\cos 0 = 1$. $\Rightarrow$ $w = F d \cos \theta$ is positive.

113. If an object slides up a frictionless inclined plane, "downward" force $mg \sin \alpha$ parallel to the inclined plane and displacement are in opposite directions.
    If $\theta = 180$, $\cos 180 = -1$. $\Rightarrow$ $w = F d \cos \theta$ is negative.

114. If a child jumps on a trampoline, the work done by the gravitational force is negative when the child is on the way up because force and displacement are in the opposite directions.

115. If a child jumps on a trampoline, the work done by the gravitational force is positive when the child is on the way down (from the maximum height) because force and displacement are in the same direction.

116. If a child jumps up and down on a trampoline, then the net work done by the gravitational force for the complete trip (coming back to the same vertical height) is zero.

117. The work done by the gravitational force for an object moving around a closed path is zero.

118. Units of work are in joules.

119. Work can be done by a specific force.

120. Work can be done by the net force.

121. 1 Joule = 1 newton × 1 meter.

122. The work done by the normal force is zero if the displacement is at 90 degrees with respect to the force.

123. A car moves to the right at constant velocity on a horizontal road. The work done by the gravitational force is zero because the force of gravity and displacement are at right angles to each other.

124. If an object slides to the right on a horizontal rough surface, then force of friction is to the left against the direction of displacement. If $\theta = 180$, $\cos 180 = -1$.
    $\Rightarrow$ $w = F d \cos \theta$ is negative, where w is the work done by friction.

125. Two masses are released in a vacuum in the outer space. The masses will accelerate towards each other immediately.

126. For the above problem, the force is not constant because the gravitational force $\propto 1 / r^2$, where r is the distance in m between their centers of the mass.

127. Unit of mechanical energy are in joules.

128. A moving object can do work at the instant it collides with another object that is initially moving left, right, or stationary.

129. A hammer can do work on a nail if it strikes it.

130. A bowling ball can do work on the pins when it strikes them.

131. Application of a non-contact force (like the gravitational force) can do work on an object.

132. When an object moves in a uniform circular motion, the work done by the centripetal force is zero, because the net force and displacement are at right angles to each other.

133. A moving object exerts a force on another only at the instant they collide if we ignore the gravitational force of attraction. Assume the masses are electrically neutral.

134. The net force $= ma$, where m is the mass in kg and a is the acceleration in m/s^2.

135. The average net force $= ma = m(\Delta v / \Delta t)$, where m is the mass in kg, a is the acceleration in m/s^2, Δv is the change in velocity in m/s, and Δt is the time interval in s.

136. The net work done *by all forces* = change in kinetic energy.

137. The net work done $=$ change in kinetic energy $= (½\, m\, v_f^{\,2}) - (½\, m\, v_i^{\,2})$, where v_f and v_i are the final and initial velocities.

138. The work-energy principle is that "the net work done is equal to the object's change in kinetic energy."

139. If there is only one force acting on an object (say gravitational force), then the *net* work done on the object by that particular force $=$ the change in its kinetic energy.

140. If an object is acted upon by two forces (say tension and weight for an elevator of mass m), then the net work done will be the sum of the work done by those individual forces (tension and gravitational force).

141. If there is no change in speed, the change in kinetic energy is zero.

142. If there is a change in speed, the change in kinetic energy is non-zero.

143. The net work done = final kinetic energy – initial kinetic energy.
The right side of this equation is positive only when the final kinetic energy is greater than the initial kinetic energy. This occurs when the final speed is greater than the initial speed.

144. Kinetic energy = $(\frac{1}{2}mv^2)$. If the mass is doubled while the speed stays constant, the kinetic energy is increased by a factor of 2.

145. Kinetic energy = $(\frac{1}{2}mv^2)$. If the velocity is doubled while the mass stays constant, the kinetic energy is increased by a factor of 4.

146. If the net work done on an object is negative, its speed decreases.

147. A hockey puck moving on a horizontal rough surface has negative work done on it by the force of friction, because the force of friction and the displacement are in the opposite directions.

148. For the above problem, the negative work reduces the speed of the hockey puck.

149. At the instant a hammer strikes a horizontal nail into a vertical wall, the force on the hammer by the nail is opposite to the displacement of the nail.

150. For the problem **149**, the force on the nail and its displacement are in the same direction.

151. For problem **149**, negative work is done on the hammer, because the force on the hammer by the nail and the direction of displacement of the nail are in the opposite directions.

152. In problem **150**, positive work is done on the nail, because the force of the hammer on the nail and the direction of displacement of the nail are in the same direction.

153. When negative work is done on a hammer, the hammer slows to rest (≈ 0 m/s) from an initial non-zero speed.

154. For the problem **149**, the decrease in kinetic energy of a hammer = work done on the nail. Ignore energy lost.

155. The total kinetic energy of two masses is their individual sum $= (\frac{1}{2} m_1 v_1^2) + (\frac{1}{2} m_2 v_2^2)$.

156. Potential energy is the energy associated with forces that depend upon the position of a body.

157. Gravitational potential energy is associated with its force, gravitational force.

158. Potential energy's dependence on position is clearly seen in freely falling objects and springs.

159. Gravitational potential energy depends on the vertical displacement of an object above a reference level.

160. If an object slides down an inclined plane from point A to point B, it does not matter where we chose the zero reference level for determining motion.

161. For problem **35**, the vertical height is h if the zero is chosen at the level of A.

162. For problem **60**, the potential energy at point C is $+mgh = mg \times$ BC if the zero level is chosen at the level of A.

163. For problem **60**, the potential energy at point A is zero if the zero level is chosen at the level of A.

164. For problem **60**, the potential energy at point A is $-mgh$ if the zero level is chosen at the level of C.

165. For problem **79**, the vertical displacement is h at point C if the zero level is chosen at the level of B.

Linear Momentum

'7

1. When does conservation of linear momentum apply?

    A.   When there is an external gravitational force acting.
    B.   When there are no external forces acting.
    C.   When there is an impulsive force acting.
    D.   When there is a frictional force acting.

2. A missile M1 flying in the air is broken into two pieces by another missile M2 fired at it. Does the law of conservation of momentum apply to just *one* of the masses (say missile M1) involved in the collision?

    A.   It does not apply to just *one* mass.
    B.   It applies to just one mass.

    For problems 3–4, choose one of the following answers as the correct response.

    A.   Linear momentum alone.
    B.   Both linear momentum and kinetic energy.
    C.   Both linear momentum and potential energy.
    D.   Both linear momentum and velocity.

3. What is conserved for an inelastic collision?

4. What is conserved for an elastic collision?

    For problems 5-6, two masses m_1 and m_2 are at distances x_1 and x_2 respectively from a point O as shown in the figure. Answer the following questions.

5. What is the distance of the center of mass of the system from point O?

    A. $(m_1 \times x_1 + m_2 \times x_2) / (m_1 + m_2)$.
    B. $(x_1 + x_2) / 2$.
    C. $(x_2 + x_1) / 2$.
    D. $(m_1 \times x_2 + m_2 \times x_1) / (m_1 + m_2)$.

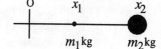

Problems 5-6

6. If u_1 and u_2 are the velocities of m_1 and m_2 respectively along the line joining them, then the velocity of the center of mass is:

    A. $(u_1 + u_2) / 2$.
    B. $(m_1 \times u_1 + m_2 \times u_2) / (m_1 + m_2)$.
    C. $(m_2 \times u_1 + m_1 \times u_2) / (m_1 + m_2)$.
    D. $(u_1 - u_2) / 2$.

7. If the two bodies collide and move with velocities v_1 and v_2 respectively, what is the velocity of the center of mass after collision?

A. $(m_1 \times v_1 + m_2 \times v_2) / (m_1 + m_2)$.
B. $(m_1 \times v_1 + m_2 \times v_2) / (m_1 - m_2)$.
C. $(v_1 + v_2) / 2$.
D. $(v_1 - v_2) / 2$.

For problems **8–9**, two point masses 5 grams and 10 grams are moving along a straight line with speeds of 8 m/s and 2 m/s respectively. Choose one of the following answers for the collision of those two masses.

A. 4 m/s in the same direction as 5 gram mass.
B. 4 m/s opposite to the motion of 5 gram mass.
C. 4/3 m/s in the same direction as 5 gram mass.
D. 4/3 m/s opposite to the motion of 5 gram mass.

8. What is the velocity of their center of mass if they move in the same direction? Choose the velocity before collision to be along positive x.

9. What is the velocity of the center of mass if they move in opposite directions? That is, 5 gram mass is moving along +x and 10 gram mass along (negative) –x.

10. During the collision of two bodies, does the center of mass of the system undergo a change of velocity?

A. Yes for elastic collisions only.
B. Yes for both elastic and inelastic collisions.
C. No for elastic collisions.
D. No for both elastic and inelastic collisions.

11. Where does conservation of linear momentum apply?

A. It applies only for elastic collisions.
B. It applies to all collisions.

For problems **12-15**, a rocket of mass M ejects fuel at the rate of m kg/s with a velocity of v m/s relative to the rocket.

12. What is the magnitude of the thrust (change of momentum of the rocket *per second*) on the rocket?

- A. $m_{\text{EJECTED MASS PER s}} \times v_{\text{VELOCITY}}$.
- B. $m_{\text{EJECTED MASS PER s}} \times v_{\text{VELOCITY}}^2$.
- C. $\frac{1}{2} \times m_{\text{EJECTED MASS PER s}} \times v_{\text{VELOCITY}}^2$.
- D. $m_{\text{EJECTED MASS PER s}} \times g_{\text{ACCELERATION DUE TO GRAVITY}}$.

13. The thrust on the rocket until the fuel is burned out:

- A. Increases.
- B. Decreases.
- C. Remains a constant.

14. What can we say about the acceleration of the rocket until the fuel is completely burned out? For simplicity, assume constant $g_{\text{ACCELERATION DUE TO GRAVITY}}$.
[Hint: Thrust T acts upwards and weight downwards. Use:
$(T)_{\text{BIGGER FORCE}} - (M \times g)_{\text{SMALLER FORCE}} = (M \times a)_{\text{NET FORCE}}$.
Here M stands for the mass of the rocket, and a for acceleration of the rocket in m/s per second.]

- A. It increases.
- B. It decreases.
- C. It is a constant.
- D. It is equal to g.

15. What is the acceleration of the rocket after the fuel has completely burned out? Assume gravity is still acting on the rocket.

- A. Zero.
- B. $-m_{\text{EJECTED MASS PER s}} \times v_{\text{VELOCITY}}$.
- C. $-g_{\text{GRAV.}}$.
- D. $g_{\text{GRAV.}}/2$.

16. How can we describe firing a bullet from a gun? Ignore the absorption of the momentum of kickback by the person or structure that holds the gun.

   A. Momentum of the bullet is in the same direction as the momentum of the gun.
   B. The vector sum of the momentum of the gun and the bullet is zero.
   C. The gun and the bullet have the same velocity.
   D. The gun and the bullet have the same momentum.

Problem 16

17. What can we say about the kinetic energy of the gun when a gun fires a bullet? Assume mass of the gun is much larger than the mass of the bullet.

   A. It is equal to that of the bullet.
   B. It is greater than that of the bullet.
   C. It is less than that of the bullet.
   D. It is zero.

18. Water jets out horizontally from a nozzle of area A with a speed v and strikes a vertical wall with nearly the same speed. What is the force exerted by water on the wall if the density of water is d and it does not rebound after striking the wall? [Hint: From a unit point of view, volume = area × thickness, and mass = volume × density.

   Mass = (area × thickness)$_{VOLUME}$ × density = [area × (speed × time)] × density. This gives us momentum

   $P_{MOMENTUM\ VECTOR} = m_{MASS}\ v_{VELOCITY} = A_{AREA} \times v_{VELOCITY}^2 \times t_{TIME} \times d_{DENSITY}.$]

Problem 18

   A. $A_{AREA}^2 \times v_{VELOCITY}^2 \times d_{DENSITY}.$
   B. $A_{AREA} \times v_{VELOCITY}^2 \times d_{DENSITY}.$
   C. $A_{AREA} \times v_{VELOCITY} \times d_{DENSITY}.$
   D. $A_{AREA} \times v_{VELOCITY}^2 \times d_{DENSITY}^2.$

19. Can change in momentum be negative?

   A. Yes.
   B. No.

20. A ball kicked to the left with a speed v, bounces off a wall and comes back with the same speed. What is the magnitude of the change in velocity?

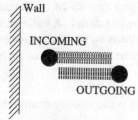

Wall

INCOMING

OUTGOING

Problem 20

    A.   Zero.
    B. v.
    C. $2v$.

For problems **21-22**, two point masses m_1 and m_2 are at a distance d meters apart.

21. What is the distance of their center of mass from mass m_1? Hint: $x_{CM} = (m_1 x_1 + m_2 x_2) / (m_1 + m_2)$. Choose x_{CM} as the origin. x_1 and x_2 are the coordinates of the masses m_1 and m_2.

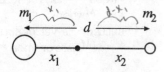

Problems 21-22

$$\frac{m_1 x_1 + m_2(d - x_1)}{m_1 + m_2}$$

    A. $(m_1 \times d) / (m_1 + m_2)$.
    B. $(m_2 \times d) / (m_1 + m_2)$.
    C. $(m_1 \times d) / (m_2)$.
    D. $(m_1 / m_2) \times d$.

22. What is the distance of their center of mass from mass m_2?

    A. $(m_1 \times d) / (m_1 + m_2)$.
    B. $(m_2 \times d) / (m_1 + m_2)$.
    C. $(m_1 \times d) / (m_2)$.
    D. $(m_1 / m_2) \times d$.

23. The earth to moon distance $= 3.8 \times 10^8$ m; radius of earth $= 6.4 \times 10^6$ m, and the earth is 80 times as massive as the moon. Where is the center of mass of the earth-moon system?

    A.   *Within* the surface of the earth between their line of centers.
    B.   *Between* the surfaces of the earth and the moon on their line of centers.
    C.   Within the surface of the moon between their line of centers.

24. The H atom and the Cl atom in an HCl molecule are 1.2×10^{-11} m apart. If chlorine's atomic weight is 35 and hydrogen's is 1, where is the molecule's center of mass?

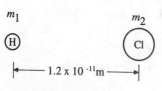

Problem 24

    A. 3.3×10^{-13} m from the Cl atom.
    B. 2.3×10^{-13} m from the H atom.
    C. 6×10^{-12} m from the H atom.
    D. 1.17×10^{-11} m from the Cl atom.

CHAPTER**seven**

For problems **25-28**, a wooden block of mass M is hung by a string of length l. A bullet of mass m is fired horizontally with a velocity v_B into the stationary block and gets embedded in it. Assume that the speed of the bullet and the block immediately after collision is v.

25. What is the velocity attained by the system v after collision?

    A. $(M_{\text{MASS OF WOOD}} \times v_{B,\text{ VEL. OF BULLET}}) / (M_{\text{WOOD}} + m_{\text{BULLET}})$.

    B. $[(M_{\text{WOOD}} + m_{\text{MASS OF BULLET}}) \times v_{B,\text{ VEL. BULLET}}] / M_{\text{WOOD}}$.

    C. $(m_{\text{BULLET}} \times v_{B,\text{ VEL. OF BULLET}}) / (M_{\text{MASS OF WOOD}} + m_{\text{BULLET}})$.

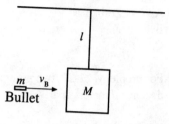

Problem 25

26. What is the speed of the bullet if the string swings through an angle θ *after impact before swinging back*? [Hint: Assume the height to which the mass gets raised $= h = l \times (1 - cos\ \theta)$].

    A. $[(M + m) / m] \times [2 \times g \times l \times (1 - cos\ \theta)]^{1/2}$.
    B. $[m / (M + m)] \times [2 \times g \times l \times (1 - cos\ \theta)]^{1/2}$.
    C. $[(M + m) / m] \times [2 \times g \times l \times (1 - cos\ \theta)]^{1/2}$.

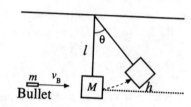

Problem 26

27. What is the loss of kinetic energy of the system?

    A. Zero.
    B. $[\frac{1}{2} \times m \times v_B^2] - [\frac{1}{2} \times (M + m) \times v^2]$.

28. The collision is:

    A. Elastic.
    B. Inelastic.

29. Conservation of linear momentum implies that with no external forces acting on the system, total momentum (the vector sum) remains a constant. Is it true for any time?

    A. Yes.
    B. No.

30. What can we say about the vector sum of momentum before and after collision of a *two*-dimensional collision between two bodies m_1 and m_2?

   A.   Momentum of m_1 is the same.
   B.   Momentum of m_2 is the same.
   C.   Vector sum of the momentum of m_1 and m_2 is the same.
   D.   Vector sum of the velocity of m_1 and m_2 is the same.

31. A system is in a stable equilibrium when its:

   A.   Kinetic energy is lowest.
   B.   Potential energy is lowest.
   C.   Kinetic energy is greatest.
   D.   Potential energy is greatest.

32. A body is in stable equilibrium when its center of mass is at the:

   A.   Lowest position.
   B.   Highest position.
   C.   Sides.

33. When a body is slightly disturbed from a position of stable equilibrium, its center of mass is:

   A.   Lowered.
   B.   Raised.
   C.   Unaltered.

34.   Two similar projectiles (assume same mass) are fired
       simultaneously as shown from the same height with the same
       magnitude for the velocity and at the same angle with respect
       to the horizontal. What can we say about the landing point
       after collision of the two masses if the collision takes place at
       the maximum height B? Make your arguments separate for
       elastic and inelastic collision.

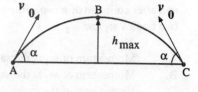

Problem 34

       A.   Elastic: The projectiles will *exchange* the path.
             Inelastic collision: Eventually they fall to earth but the
             one on the left will be found to the left of the
             maximum height.
       B.   For both elastic and inelastic collision, they fall
             vertically down.
       C.   After the collision, they go even higher.
       D.   Elastic: The projectiles will retrace the path. Inelastic
             collision: They fall to earth vertically down from the
             maximum height with no horizontal component for
             velocity.

35.   Two similar billiard balls move towards each other with equal
       speed. Do they both have the same momentum?

       D.   Yes.
       E.   No.

36.   What is the unit of momentum?

       A. $kg \cdot m$.
       B. $kg \cdot m^2$.
       C. $kg \cdot m / s$.
       D. $kg \cdot m / s^2$.

37.   You are given two bullets, A and B, of equal mass. Bullet A
       bounces back from a block of wood with its speed unchanged.
       Bullet B passes through with little resistance. Which of these
       can be used to knock down a piece of wood?

       A.   Bullet A.
       B.   Bullet B.
       C.   It does not matter which bullet we use.

38.   If you are stranded on a frictionless lake, what can you do to
       go east?

       A.   Remove your clothes and throw east.
       B.   Remove your clothes and throw west.

39. Which of the following formula is wrong?

    A. $F_{net} = \Delta p \,/\, \Delta t$.
    B. $F_{net} \times \Delta t = \Delta p$.
    C. $F_{net} = ma$.
    D. $F_{net} = m \times (\Delta v \,/\, \Delta t)$. (ma)
    E. $F_{net} = \Delta p \times \Delta t$.

40. A ball reaches a racket at 60 m/s along **+x**, and leaves the racket in the opposite direction with the same speed. Assume mass of the ball is 0.05 kg and that the contact time is 0.02 seconds. Find the force exerted by the racket on the ball.

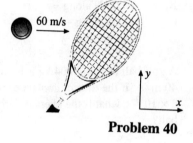

Problem 40

    A.   300 N along **+x**.
    B.   300 N along **−x**.
    C.   0 N.

41. What is comparable to a 300−N force?

    A.   We can lift a 300−kg person off the ground. Wow!
    B.   We can lift a 30−kg child off the ground.

42. What is the contact time of a bat if a ball with an approach velocity of 30 m/s leaves the bat at 60 m/s in the opposite direction. Assume a force of 400 N and that the mass is 0.06 kg.
[Hint: Is the magnitude of the change in velocity 30 or 90 m/s?]

    A.   0.0015 s
    B.   0.0025 s.
    C.   0.0135 s.
    D.   0.0045 s.

43. If the net force acting on an object is zero, what can we conclude?

    A.   It is moving at constant speed.
    B.   It is decelerating.
    C.   It has constant momentum.

44. If you drop a fish of mass 2 kg at rest from a height of 2 m into a net, what is the average force exerted by the net on the fish as the fish comes to rest? Assume a contact time of 0.7 seconds. Assume **+y** is up. Ignore gravitational potential energy after the fish hits the net.

Problem 44

A. 14.05 N along + *y*.
B. 14.05 N along − *y*.
C. 18.05 along + *y*.
D. 18.05 along − *y*.

45. A golf ball of mass 0.04 kg is hit to the right with a speed of 40 m/s. If the time of contact of the ball with the golf club is 2×10^{-3} s, what is the average force exerted by the club on the ball?

A. 200 N.
B. 300 N.
C. 400 N.
D. 800 N.

46. What is impulse?

A. It is force.
B. It is the change in velocity.
C. It is momentum.
D. It is the change in momentum.

47. While we catch a baseball, we catch it at the front and make the hands ride with the ball backward. Why is that?

A. To reduce impulse.
B. To increase the time of contact.

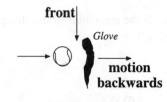

Problem 47

48. For a karate kid, what is more important to break a brick?

A. Increase the time of contact.
B. Decrease the time of contact.

49. What can be done to to reduce the impact force when a plate falls on the ground?

A. Spill water on the floor.
B. Reduce the time of contact between the plate and the floor.
C. Increase the hardness of the floor.
D. Make the carpets as padded as possible.

---

50. Boxing gloves are filled with soft material to:

    A.   Increase the time of contact.
    B.   Reduce the time of contact.

51. A person of mass 70 kg jumps to the ground and lands with a 20,000-N force. What is the velocity with which he/she lands? Assume a time of contact 0.04 s and that the velocity of the person is zero *after* landing.

    A.   5.1 m/s.
    B.   7.3 m/s.
    C.   9.5 m/s.
    D.   11.4 m/s.

52. For problem **51**, calculate the height of fall. Assume all potential energy at the beginning.

    A.   3.5 m.
    B.   4.5 m.
    C.   5.5 m.
    D.   6.5 m.

For problems **53–57**, a mass m_1 moves with a constant velocity v_{1i} to the right. It then collides with another mass m_2 at rest. After collision, they stick together and move with velocity v. Choose one of the following answers.

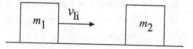

    A.   Net force is zero.
    B.   Net force is to the right.
    C.   Net force is to the left.

Problems 53-57

53. What is the direction of the net force on mass m_1 before collision with mass m_2?

54. What is the direction of the net force on mass m_2 before collision with mass m_1?

55. At the instant they collide, what is the direction of the net force on mass m_1?

56. At the instant they collide, what is the direction of the net force on mass m_2?

57. If you consider m_1 and m_2 as a system, what is the net force on that system at the instant they collide?

58. For problems **58–59**, a 10,000 kg car moving at 12 m/s collides with a similar car at rest. After collision, they move together. What is the velocity after collision?

    A.   12 m/s.
    B.   120 m/s.
    C.   6 m/s.
    D.   60 m/s.
    E.   0 m/s.

59. In the above problem, what happens if they have a head on (totally inelastic) collision? That is, they move towards each other with a speed of 12 m/s before collision.

60. [Use the equation $m_1v_{1i} + m_2v_{2i} = m_1v_{1f} + m_2v_{2f}$]. Assume a head on collision between two football players. $m_1 = 120$ kg, $m_2 = 75$ kg, $v_{1i} = 3$ m/s, and $v_{2i} = -3$ m/s. If the two masses stick together after collision, what is the velocity of the two together?

    A.   0.69 m/s in the original direction of the bigger person.
    B.   0.69 m/s in the original direction of the smaller person.
    C.   2.49 m/s in the original direction of the bigger person.
    D.   2.49 m/s in the original direction of the smaller person.

61. A mass m_1 (= 2 kg) is placed on a smooth inclined plane and released at rest. Another mass m_2 (= 4 kg) is placed at the bottom of the same inclined plane. What is the speed of the mass m_1 at the bottom just before it collides with the mass m_2? Assume 30 degrees angle for the inclined plane and a separation of 12 m between m_1 and m_2.

    A.   10.95 m/s.
    B.   14.45 m/s.

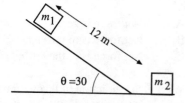

Problem 61

For problem **61**, m_1 collides with mass m_2 at the base of the inclined plane. Assume the two blocks stick together after the collision. Find the speed of the two together after collision for the following situations (problems **62–64**) from the choices listed below.

A.   3.65 m/s to the right.
B.   3.65 m/s to the left.
C.   1.65 m/s to the right.
D.   1.65 m/s to the left.
E.   0.35 m/s to the left.

62.   When the mass m_2 is initially at rest.

63.   When the mass m_2 is initially moving to the left at 3 m/s at the base.

64.   When the mass m_2 is initially moving to the left at 6 m/s at the base.

65.   For the two masses (after collision) in problem **64**, how high do they rise from ground level?

A.   0.0061 m.
B.   3 m.
C.   12 m.
D.   101 m.

For problems **66–74**, a mass m_2 (= 2 kg) is at rest at the edge of a smooth horizontal tall table. Another mass m_1 (= 3 kg) placed to the left of m_2 slides to the right at a constant velocity of 5 m/s and collides with m_2. After collision they stick together. The table is 4 m high.

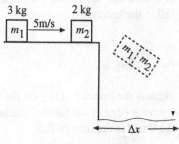

Problems 66-74

66.   What is the horizontal speed of m_2 immediately after collision?

A.   1 m/s.
B.   2 m/s.
C.   3 m/s.
D.   4 m/s.
E.   Zero.

67.   What is the vertical speed of m_2 immediately after collision?

A.   Zero.
B.   Non-zero.

68. What is the time taken by the mass m_2 to reach the ground?

    A.   0.64 s.
    B.   0.89 s.

69. What is the horizontal speed of m_1 immediately after collision?

    A.   Zero.
    B.   Non-zero.

70. What is the horizontal displacement for the two masses from the edge of the table to the landing point?

    A. $\Delta x = 0.67$ m.
    B. $\Delta x = 1.67$ m.
    C. $\Delta x = 2.67$ m.

71. Is momentum conserved along the vertical at the start immediately after collision?

    A.   Yes.
    B.   No.

72. After collision, does the momentum change for the two masses during fall?

    A.   Yes.
    B.   No.

73. Is there a net force along the horizontal after collision as they fall to the ground?

    A.   Yes.
    B.   No.

74. What is the impulse acting on the two masses after collision from the edge of the table to the landing point of the masses? Assume t is the time of fall.

    A. $m_1 g \times t$.
    B. $m_2 g \times t$.
    C. $[(m_1 + m_2)g] \times t$.
    D.   0.

For problems **75-78**, a bullet (mass of the bullet = 200 grams = 0.2 kg) is fired at a block (mass of the block = 2 kg) that is hung from a ceiling. The velocity of the bullet before collision = 400 m/s. Choose one of the following answers:

A. 32 m/s.
B. 38.2 m/s.
C. 34.55 m/s.
D. 36.4 m/s.
E. 43 m/s.

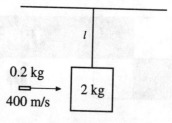

Problems 75-78

75.   After collision, if the bullet is embedded in the block, find the velocity of the block and the bullet after collision. Assume the block was at rest before collision.

76.   If the block is swinging at the bottom point to the left at 2 m/s just before impact, and the bullet is traveling to the right, find the velocity of the bullet and the block after collision. Assume the bullet is embedded in the block.

77.   If the block is swinging to the right at the bottom point at 2 m/s, find the velocity of the bullet and the block after collision. Assume the bullet is embedded in the block.

78.   If the bullet passes through the block and exits at 80 m/s, find the velocity of the block after collision. Assume that the block was at rest before collision.

TRUE?   **CONCEPT CHECK**

m u s t K n O w

79.   Linear momentum is the product of mass m and its velocity v.

80.   If mass m is a constant, the change in momentum $\Delta p = m\Delta v$, where Δp is the change in momentum and Δv is the change in velocity ($= v_f - v_i$).

81.   For small time intervals, $\Delta p / \Delta t = (m\Delta v)/ \Delta t.$ $\Rightarrow \Delta p / \Delta t$ $= ma =$ the net force.

82. Δp, the change in momentum of an object ($= m\Delta v$ if mass is a constant), is a vector.

83. Δp, the change in momentum of an object is net force $\times \Delta t$.

84.   The change in momentum is called impulse.

85.   If there are two trucks of the same mass, the one going at a higher speed has a larger (magnitude of) momentum.

86.   The more momentum an object has, the harder it is to stop.

87.   The more momentum an object has, the larger the braking force to stop it, in a given distance.

88.   A non-zero net force on an object changes the momentum.

89.   For an object moving in uniform circular motion at constant speed, the momentum vector is always changing because the velocity vector is continuously changing its direction.

90.   Change in momentum can be negative.

91.   When two balls collide, the only significant force is the force they exert on one another at the instant they collide if we ignore gravitational, electrical, and magnetic forces.

92. The vector sum of the momentum of two masses m_1 and m_2 is $m_1 v_1 + m_2 v_2$ where v_1 and v_2 are the velocities (not the speed) of the masses m_1 and m_2.

93. A firecracker at rest on the ground is about to explode. It has zero momentum before it explodes.

94. The vector sum of a duck (at rest) that has just exploded inside an oven is zero.

95. The vector sum of the momenta of all bodies before collision is equal to the vector sum of the momenta after collision.

96. A ball falls toward the ground. If the system is defined as just the ball, then gravitational force accelerates the ball (in the absence of friction) and its speed continuously increases.

97. A ball falls towards the ground. If the system is defined as just the ball, then momentum is not conserved.

98. A ball falls towards the ground. If the system is defined as just the *earth*, then momentum is not conserved.

99. A ball falls towards the ground. If the *earth* and the *ball* are defined as the system, then momentum is conserved.

100. If two objects collide and if they are very elastic (like billiard balls), then kinetic energy is also conserved.

101. In elastic collisions, kinetic energy is conserved.

102. In inelastic collisions, kinetic energy is not conserved.

103. For any elastic head-on collision, the relative speed of the two masses after collision is the same as before no matter how big or small the masses are.

104. A billiard ball of mass m_1 and speed v_1 is shot at another ball of mass m_2 at rest. If they undergo elastic collision, then m_1 comes to a stop and m_2 moves with speed v if and only if the two masses are the same and collision is head on (the velocity vector and relative position vector are collinear – no glancing collisions).

Choose one of the following as your answer for problems
1-14, with the axis of rotation at point A. Assume that the
forces are acting on a rod of negligible mass. .

A.   Net torque $= I \times \alpha = F \times R$ clockwise.
B.   Net torque $= I \times \alpha = F \times R$ counterclockwise.
C.   Net torque $= I \times \alpha = 0. \Rightarrow \alpha = 0.$
D.   Net torque $= I \times \alpha = F \times R \times sin$ 30 clockwise.
E.   Net torque $= I \times \alpha = F \times R \times sin$ 30 counterclockwise.

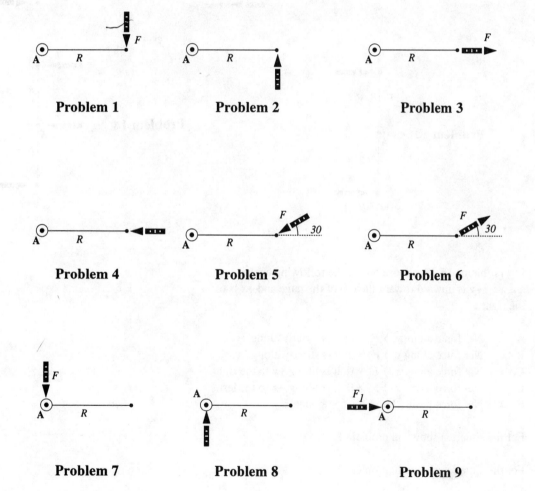

Problem 1   **Problem 2**   **Problem 3**

Problem 4   **Problem 5**   **Problem 6**

Problem 7   **Problem 8**   **Problem 9**

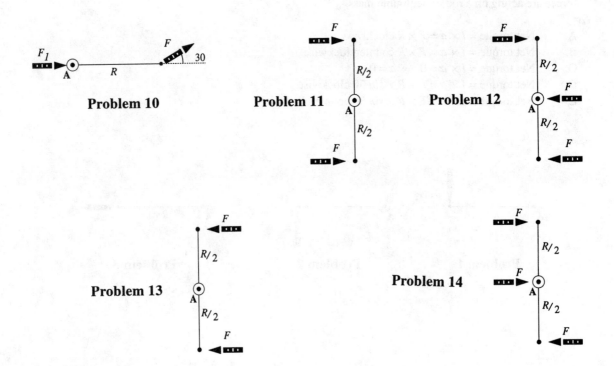

Problem 10

Problem 11

Problem 12

Problem 13

Problem 14

For problems **15-28**, choose one of the following as the
answer. **+y** is upward toward the top of the page and **+x** is to
the right.

A. Net force along y, $\sum F_y \neq 0$; it is strictly along **+y**.
B. Net force along y, $\sum F_y \neq 0$; it is strictly along **−y**.
C. Net force along x, $\sum F_x \neq 0$; it is along **+x** to the right.
D. Net force along x, $\sum F_x \neq 0$; it is along **−x** to the left.
E. Net force has both x and y components.

15. For the situation shown in problem **1**.

16. For the situation shown in problem **2**.

17. For the situation shown in problem **3**.

18. For the situation shown in problem **4**.

19. For the situation shown in problem **5**.

20. For the situation shown in problem **6**.

21. For the situation shown in problem **7**.

22. For the situation shown in problem **8**.

23. For the situation shown in problem **9**.

24. For the situation shown in problem **10**.

25. For the situation shown in problem **11**.

26. For the situation shown in problem **12**.

27. For the situation shown in problem **13**.

28. For the situation shown in problem **14**.

Answer the questions **29-31** based on this figure.

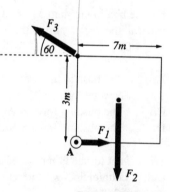

Problems 29-31

29. The vector sum of the horizontal force components, $\sum F_x = 0$, gives us:

   A. $F_3 = F_1 + F_2$.
   B. $F_3 \times sin\ 60 = F_2$.
   C. $F_3 \times cos\ 60 = F_2$.
   D. $F_3 \times cos\ 60 = F_1$.
   E. $F_3 \times sin\ 60 = F_1$.

30. The vector sum of the vertical force components, $\sum F_y = 0$, gives us:

   A. $F_3 = F_1 + F_2$.
   B. $F_3 \times sin\ 60 = F_2$.
   C. $F_3 \times cos\ 60 = F_2$.
   D. $F_3 \times cos\ 60 = F_1$.
   E. $F_3 \times sin\ 60 = F_1$.

31. The sum of all the clockwise torques = the sum of all the counterclockwise torques. $\sum \tau = 0$ gives us:

   A. $F_3 \times 3 = F_2 \times 3.5$.
   B. $F_3 \times 3 = F_2 \times 3.5 \times cos\ 60$.
   C. $F_3 \times cos\ 60 \times 3 = F_2 \times 3.5$
   D. $(F_3 \times sin\ 60 \times 3) + (F_3 \times cos\ 60 \times 3) = F_2 \times 3.5$.
   E. $F_3 \times cos\ 60 \times 3 = (F_1 \times 3.5) + (F_2 \times 7)$

CHAPTEReight

32. A drinking glass is lying on a table. How can we make the glass stand upright?

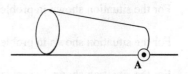

Problem 32

    A. Apply no net force.
    B. Create a net torque of zero.
    C. Create net torque clockwise about point A.
    D. Create net torque counterclockwise about point A.

A box is about to tip over for problems **33-35**.

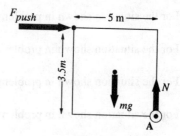

33. What happens if the vector sum of the horizontal force components (the net force along the horizontal) $\sum F_x = 0$?

    A. The push to the right = Force of friction to the left.
    B. Normal force = Weight.
    C. The pushing force = Weight.
    D. The pushing force = Normal force.

Problems 33-35

34. For the box in problem **33**, what happens if net force along the vertical, the vector sum of the vertical force components $\sum F_y = 0$?

    A. The push to the right = Force of friction to the left.
    B. Normal force = Weight.
    C. The pushing force = Weight.
    D. The pushing force = Normal force.

35. What if net torque is zero (the sum of clockwise torques = the sum of counterclockwise torques) about point A for the box in problem **33**?

    A. $m \times g \times 5 = F_{push} \times 3.5$.
    B. $m \times g \times 2.5 = F_{push} \times 5$.
    C. $N \times 2.5 = m \times g \times 3.5$.
    D. $(N \times 2.5) + (F_{push} \times 3.5) = m \times g \times 2.5$.
    E. $m \times g \times 2.5 = F_{push} \times 3.5$.

36. A hanging rectangular frame is in equilibrium. What happens if the axis of rotation is at point A?
(Hint: Net torque is zero about point A.)

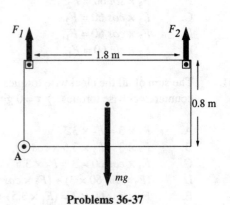

    A. $m \times g \times 1.8 = F_2 \times 0.9$.
    B. $(F_1 \times 1.8) + (F_2 \times 1.8) = m \times g \times 0.9$.
    C. $m \times g \times 0.9 = F_2 \times 1.8$.
    D. $m \times g \times 1.8 = F_1 \times 0.9$.
    E. $(F_2 \times 1.8) + (F_1 \times 1.8) = m \times g \times 0.4$.

Problems 36-37

37. For the frame in problem **36**, what if the vector sum of the vertical force components $\sum F_y = 0$?

    A. $F_1 = F_2 + (m \times g)$
    B. $F_2 = F_1 + (m \times g)$
    C. $F_1 + F_2 = m \times g$.

38. For a two-dimensional car as shown, assume uniform mass distribution and that the axis of rotation is at point A. Assume weight acts halfway between N_R (the normal force at the rear tire) and N_F (the normal force at the front tire). What if net force (the vector sum of the vertical force components) along y is zero?

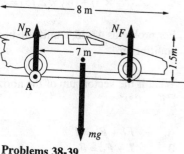

    A. $N_R = N_F + (m \times g)$.
    B. $N_R + N_F = m \times g$.
    C. $N_F = N_R + (m \times g)$.

Problems 38-39

39. What if net torque about point A is zero in problem **38**? (Hint: Sum of the clockwise torques = sum of the counterclockwise torques.)

    A. $N_F \times 8 = m \times g \times 3.5$.
    B. $(N_F \times 7) + (N_R \times 7) = m \times g \times 3.5$.
    C. $(N_F \times 15) + (N_R \times 1.5) = m \times g \times 3.5$.
    D. $N_F \times 1.5 = m \times g \times 3.5$.
    E. $N_F \times 7 = m \times g \times 3.5$.

For problems **40-42**, you load groceries into a van where W_2 is the weight of the groceries.

40. If net force (the vector sum of the vertical force components) along y is zero, what is correct?

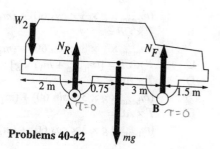

    A. $W_2 + N_R = (m \times g) + N_F$.
    B. $W_2 + N_F = (m \times g) + N_R$.
    C. $W_2 + (m \times g) = N_R + N_F$.
    D. $W_2 = N_R + N_F$.

Problems 40-42

41. If net torque about point A is zero, what is true? (Note: Sum of the clockwise torques = sum of the counterclockwise torques.)

    A. $(W_2 \times 2) + (N_F \times 3.75) = m \times g \times 0.75$.
    B. $(W_2 \times 2) + (m \times g \times 0.75) = N \times 3.75$.
    C. $W_2 \times 2 = (N_F \times 3.75) + (m \times g \times 0.75)$.
    D. $(W_2 \times 2) + (N_R \times 2) + (m \times g \times 0.75) + (N_F \times 1.5) = 0$.

42. For problem **41**, if net torque is zero about an axis passing through point B, what happens?

    A. $(m \times g \times 3) + (w_2 \times 2) = N_R \times 3.75$.
    B. $(m \times g \times 3) + (w_2 \times 5.75) = N_R \times 3.75$.
    C. $(N_R \times 3) + (w_2 \times 2.75) = m \times g \times 0.75$.

    For problems **43-45**, a uniform beam 6 m long is in static equilibrium.

43. What if the vector sum of the horizontal components $\sum F_x = 0$?

    A. $F_V + F_H = T$.
    B. $T + F_V = F_H$.
    C. $T = F_H$.

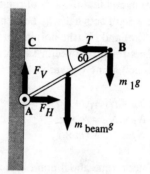

Problems 43-45

44. What if the vector sum of the vertical force components $\sum F_y = 0$?

    A. $F_V = T + (m_1 \times g) + (m_{beam} \times g) + F_H$.
    B. $F_V = (m_{beam} \times g) + (m_1 \times g)$.
    C. $m_{beam} \times g = m_1 \times g$.

45. What if net torque is zero about point A?
    (Sum of the clockwise torques about the axis of rotation = sum of the counterclockwise torques about the same axis of rotation.)

    A. $T \times CB =$
       $(m_{beam} \times g \times 3 \times cos\ 60) + (m_1 \times g \times 6 \times cos\ 60)$.
    B. $T \times CB =$
       $(m_{beam} \times g \times 3 \times cos\ 60) + (m_1 \times g \times 6 \times sin\ 60)$.
    C. $T \times CB = [(m_{beam} + m_1) \times g] \times 3 \times cos\ 60$.
    D. $T \times AC =$
       $(m_{beam} \times g \times 3 \times sin\ 60) + (m_1 \times g \times 6 \times sin\ 60)$.
    E. $T \times AC =$
       $(m_{beam} \times g \times 3 \times cos\ 60) + (m_1 \times g \times 6 \times cos\ 60)$.

For problems **46-48**, a uniform beam 4 meters long is in equilibrium.

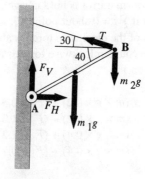

Problems 46-48

46. What if the vector sum of the horizontal force components $\sum F_x = 0$?

A. $T \times cos\ 30 = F_H.$
B. $T \times sin\ 30 = F_H.$
C. $T \times cos\ 60 = F_H.$
D. $T \times sin\ 60 = F_H.$

47. What if the vector sum of the vertical force components is $\sum F_y = 0$?

A. $(T \times sin\ 30) + F_V = (m_1 \times g) + (m_2 \times g).$
B. $(T \times cos\ 30) + F_V = (m_1 \times g) + (m_2 \times g).$

48. What if $\sum \tau = 0$ about point A?
(Sum of the clockwise torques about the axis of rotation = sum of the counterclockwise torques about the same axis of rotation.)

A. $[(m_1 \times g) + (m_2 \times g)] \times 2 \times cos\ 40 = [(T \times sin\ 30) + (T \times cos\ 30)] \times 4 \times sin\ 40.$
B. $T \times sin\ 30 \times 4 \times cos\ 40 = m_2 \times g \times 4 \times cos\ 40.$
C. $T \times sin\ 30 \times 4 \times sin\ 40 = (m_2 \times g \times 2 \times cos\ 40) + (m_1 \times g \times 2 \times cos\ 40).$
D. $(T \times sin\ 30 \times 4 \times cos\ 40) + (T \times cos\ 30 \times 4 \times sin\ 40) = (m_2 \times g \times 4 \times cos\ 40) + (m_1 \times g \times 2 \times cos\ 40).$
E. $(T \times sin\ 30 \times 4 \times sin\ 40) + (T \times cos\ 30 \times 4 \times cos\ 60) = (m_2 \times g \times 4 \times sin\ 40) + (m_1 \times g \times 2 \times sin\ 40).$

49. A child of mass m_2 plays seesaw alone (part of the seesaw is shown in the figure). What happens if $\sum \tau = 0$ about an axis passing through the fulcrum (support).
(Sum of the clockwise torques about the axis of rotation = sum of the counterclockwise torques about the same axis of rotation.)

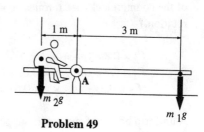

Problem 49

A. $m_2 \times g \times 1 = m_1 \times g \times 4.$
B. $m_2 \times g \times 1 = m_1 \times g \times 3.$
C. $m_2 \times g \times 4 = m_1 \times g \times 3.$

50. A uniform mirror is held upright, without tilting, by supports. What if $\sum \tau = 0$ about point A?
(Sum of the clockwise torques about the axis of rotation = sum of the counterclockwise torques about the same axis of rotation.)

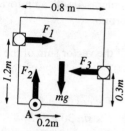

Problem 50

- A. $(m \times g \times 0.2) + (F_3 \times 0.3) = F_1 \times 1.2$.
- B. $(m \times g \times 0.2) + (F_1 \times 1.2) = F_3 \times 0.3$.
- C. $(m \times g \times 0.4) + (F_1 \times 1.2) = F_3 \times 0.3$.
- D. $(m \times g \times 0.4) + (F_2 \times 0.2) = (F_3 \times 0.3) + (F_1 \times 1.2)$.

For problems **51-53**, a diving board of negligible mass holds two people of masses m_1 and m_2. The supports provide forces F_1 and F_2 as shown.

51. What if the vector sum of the vertical force components $\sum F_y = 0$?

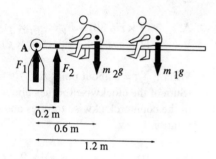

Problems 51-53

- A. $F_1 + (m_1 \times g) = F_2 + (m_2 \times g)$.
- B. $F_2 = (m_2 \times g) + (m_1 \times g)$.
- C. $F_1 + F_2 = m_2 \times g$.
- D. $F_1 + F_2 = (m_2 \times g) + (m_1 \times g)$.

52. What is the torque of the force F_1 about point A?

- A. Zero.
- B. Non-zero.

53. What if net torque about A is zero?
(Sum of the clockwise torques about the axis of rotation = sum of the counterclockwise torques about the same axis of rotation.)

- A. $F_2 \times 0.8 = (m_1 + m_2) \times 0.6$.
- B. $F_2 \times 0.2 = (m_2 \times g \times 0.6) + (m_1 \times g \times 1.2)$.
- C. $(F_2 \times 0.2) + (m_1 \times g \times 0.6) = m_2 \times g \times 1.2$.

54. A uniform bridge of mass m is supported as shown on two edges. What if net torque is zero about point A?
(Sum of the clockwise torques about the axis of rotation = sum of the counterclockwise torques about the same axis of rotation.)

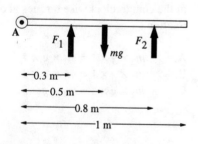

Problem 54

- A. $F_1 + F_2 = m \times g \times 0.5$.
- B. $(F_1 \times 0.5) + (F_2 \times 0.8) = m \times g \times 0.5$.
- C. $(F_1 \times 0.3) + (F_2 \times 0.8) = m \times g \times 0.5$.
- D. $(F_1 + F_2) \times 0.3 = m \times g \times 0.5$.

For problems **55-57**, a volleyball net is secured to the ground.

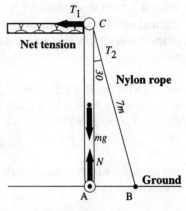

Problems 55-57

55. What if the vector sum of the horizontal force components $\sum F_x = 0$?

A. $T_1 = T_2$.
B. $T_1 = T_2 \times cos\ 30$.
C. $T_1 = T_2 \times sin\ 30$.
D. $(m \times g) + T_2 = T_1$.

56. What if $\sum \tau = 0$ about point A?
(Sum of the clockwise torques about the axis of rotation = sum of the counterclockwise torques about the same axis of rotation.)

A. $[T_1 + (m_1 \times g)] \times 7 \times cos\ 30 = T_2 \times 7$.
B. $T_1 \times cos\ 30 \times 7 \times cos\ 30 = T_2 \times 7$.
C. $T_1 \times 7 \times cos\ 30 = T_2 \times sin\ 30 \times 7 \times cos\ 30$.
D. $m \times g \times [(7 \times cos\ 30)/\ 2] = T_2 \times 7 \times sin\ 30$.
E. $N \times 7 \times sin\ 30 = (m \times g \times 7) + (T_2 \times cos\ 30 \times 3.5)$.

57. What if the vector sum of the vertical force components $\sum F_y = 0$?

A. $T_1 = T_2$.
B. $m \times g = N$.
C. $(m \times g) + (T_2 \times cos\ 30) = N$.
D. $(m \times g) + (T_2 \times sin\ 30) = N$.
E. $N = m \times g$.

For problems **58-60**, a sign hangs from a pole.

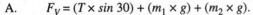

Problems 58-60

58. What if the vector sum of the horizontal force components $\sum F_x = 0$?

A. $T \times cos\ 30 = F_P$.
B. $T \times sin\ 30 = F_P$.
C. $T = F_P$.

59. What if the vector sum of the vertical force components $\sum F_y = 0$?

A. $F_V = (T \times sin\ 30) + (m_1 \times g) + (m_2 \times g)$.
B. $F_V + (T \times sin\ 30) = (m_1 \times g) + (m_2 \times g)$.
C. $F_V + (T \times cos\ 30) = (m_1 \times g) + (m_2 \times g)$.

CHAPTER**eight**

60. What if $\sum \tau = 0$ about point A?
(Sum of the clockwise torques about the axis of rotation = sum
of the counterclockwise torques about the same axis of
rotation.)

A. $(T \times sin\ 30 \times 0.5) + (m_1 \times g \times 0.8) = (m_2 \times g \times 0.7)$
B. $m_1 \times g \times 0.8 = m_2 \times g \times 0.7.$
C. $T \times cos\ 30 \times 0.5 = (m_1 \times g \times 1.3) + (m_2 \times g \times 2).$
D. $T \times sin\ 30 \times 0.5 = (m_1 \times g \times 1.3) + (m_2 \times g \times 2).$

For problems **61-63**, a drum is held steady about an axis
passing through point A.

61. What if $\tau_{net} = 0$ about an axis passing through point A?
(Sum of clockwise torques about the axis of rotation = sum of
counterclockwise torques about the same axis of rotation.)

A. $T_1 \times R = T_2 \times R.$
B. $(T_1 + T_2) \times R = T_1 \times R.$
C. $T_2 = 2 \times T_1.$

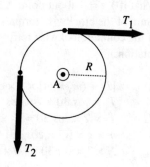

62. If net torque is clockwise, which tension is greater?

A. $T_1 \rangle T_2.$
B. $T_1 = T_2.$
C. $T_1 \langle T_2.$

Problems 61-63

63. If net torque is counterclockwise, which tension is greater?

A. $T_1 \rangle T_2.$
B. $T_2 \rangle T_1.$
C. $T_1 = T_2.$

64. An object has two drums with radii R_1 and R_2 and is acted on
by two forces. What can we conclude?

A. Net torque is clockwise.
B. Net torque is counterclockwise.
C. Net torque is zero.

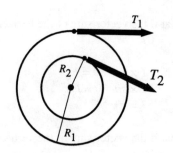

Problem 64

65. A basketball is held steady on the fingertip. If there is a force, *F* that acts as shown, what can we conclude? Assume *F* is perpendicular to this page. Assume the axis of rotation passes through the finger.

    A.   Net torque about the fingertip is clockwise.
    B.   Net torque about the fingertip is counterclockwise.
    C.   Net torque about the fingertip is zero.

66. For problem **65**, what is the contribution of normal force (exerted by the finger on the basketball) towards the torque?

    A.   Zero.
    B.   Non-zero.

67. For problem **65**, if weight does not act vertically above the fingertip, what can we say about the net torque about an axis passing through the fingertip?

    A.   There is a net unbalanced torque that will rotate the basketball and make it fall.
    B.   Net torque will be zero.

68. A cylinder of mass *m* rotates about an axis passing through its center in response to a hanging weight as shown. What is the net torque on the cylinder? Ignore frictional torque.

    A. $T \times R$ clockwise.
    B. $T \times R$ counterclockwise.
    C. $m_1 \times g \times R$ clockwise.
    D. $m_1 \times g \times R$ counterclockwise.

For problems **69-70**, a beam of mass *m* is held steady.

69. What if $\tau_{net} = 0$ about point A? (Assume *sin* 30 = *cos* 60). (Sum of the clockwise torques about the axis of rotation = sum of the counterclockwise torques about the same axis of rotation.)

    A. $T \times 3 = m \times g \times 3$.
    B. $T \times 6 = m \times g \times 3$.
    C. $[T + (F_1 \times cos\ 60)] \times 3 = m \times g \times 6$.
    D. $[T \times (F_1 \times sin\ 60)] \times 6 = m \times g \times 3$.

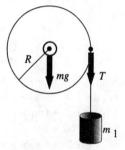

Problem 68

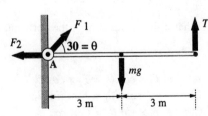

Problems 69-70

70. What if the vector sum of the horizontal force components $\sum F_x = 0$?

A. $F_1 \times cos\ \theta = T.$
B. $F_1 \times sin\ \theta = T.$
C. $m \times g = T.$
D. $F_1 \times cos\ \theta = F_2.$

For problems **71-72**, a uniform gate of mass m is hung as shown with a rope on two hinges, 1.5 m apart. Hinge 1 is directly above hinge 2.

71. What if the vector sum of the horizontal components is $\sum F_x = 0$?

A. $F_{V1} = F_{V2}.$
B. $F_{V1} + F_{V2} = T \times cos\ 60.$
C. $F_{V1} + F_{V2} = T \times sin\ 60.$

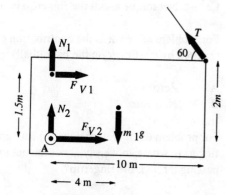

Problems 71-72

72. What if net torque = 0 (about an axis passing through point A)?
(Sum of the clockwise torques about the axis of rotation = sum of the counterclockwise torques about the same axis of rotation.)

A. $(F_{V1} \times 1.5) + (N_1 \times 0) + (m_1 \times g \times 4) =$ $(T \times sin\ 60 \times 10) + (T \times cos\ 60 \times 2).$
B. $(F_{V1} \times 1.5) + (m_1 \times g \times 4) = T \times sin\ 60 \times 10.$
C. $(F_{V1} \times 1.5) + (m_1 \times g \times 4) = T \times cos\ 60 \times 10.$
D. $(F_{V1} \times 1.5) + (m_1 \times g \times 4) = (T \times cos\ 60 \times 10) +$ $(T \times sin\ 60 \times 2).$
E. $(F_{V2} \times 1.5) + (m_1 \times g \times 4) = (T \times cos\ 60 \times 10) +$ $(T \times sin\ 60 \times 2).$

For problems **73-75**, a person of mass m_1 stands free on a bar of mass m_2. Tension in wire 1 makes an angle θ_1 with the horizontal, and tension in wire 2 makes an angle θ_2 with the horizontal.

73. What if the vector sum of the horizontal force components $\sum F_x = 0$?

A. $T_1 \times cos\ \theta_1 = T_2 \times cos\ \theta_2.$
B. $T_1 \times cos\ \theta_1 = T_2 \times sin\ \theta_2.$
C. $T_1 \times sin\ \theta_1 = T_2 \times cos\ \theta_2.$

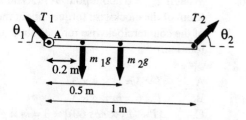

Problems 73-75

74.   What if $\sum \tau = 0$ about point A?
        (Sum of the clockwise torques about the axis of rotation = sum
        of the counterclockwise torques about the same axis of
        rotation.)

        A. $(m_1 \times g \times 0.2) + (m_2 \times g \times 0.5) = T_2 \times sin\ \theta_2 \times 1.$
        B. $(m_1 \times g \times 0.2) + (m_2 \times g \times 0.5) = T_2 \times cos\ \theta_2 \times 1.$
        C. $(m_1 \times g \times 0.2) + (m_2 \times g \times 0.5) = (T_2 \times cos\ \theta_2 \times 1) +$
                $(T_2 \times sin\ \theta_2 \times 1).$

75.   What if the vector sum of the vertical force components
        $\sum F_y = 0$?

        A. $(m_1 + m_2) \times g = T_1 + T_2.$
        B. $m_1 + m_2 = T_1 + T_2.$
        C. $T_1 \times sin\ \theta_1 = T_2 \times sin\ \theta_2.$
        D. $(T_1 \times sin\ \theta_1) + (T_2 \times sin\ \theta_2) = (m_1 \times g) + (m_2 \times g).$

        For problems **76-78**, a ladder stands against a frictionless wall.
        The base of the ladder makes a 60° angle with the horizontal.

76.   What if the vector sum of the horizontal force components is
        $\sum F_x = 0$?

        A. $F_{wall} = \mu.$
        B. $F_{wall} = F_{friction}.$
        C. $F_{wall} = m \times g.$

77.   What if the vector sum of the vertical force components
        $\sum F_y = 0$?

        A. $N = \mu \times F_{friction}.$
        B. $N + W = F_{friction}.$
        C. $N = W.$

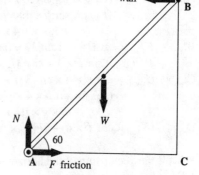

Problems 76-78

78.   What if $\sum \tau = 0$? (Length of the ladder is given as AB.)
        (Sum of the clockwise torques about the axis of rotation = sum
        of the counterclockwise torques about the same axis of
        rotation.)

        A. $W \times (AB\ /\ 2) = (N \times AC) + (F_{wall} \times AB).$
        B. $W \times [(AB \times sin\ 60)\ /\ 2] = F_{wall} \times AB \times cos\ 60.$
        C. $W \times AB \times cos\ 60 = F_{wall} \times AB.$
        D. $W \times [(AB \times cos\ 60)\ /\ 2] = F_{wall} \times AB \times sin\ 60.$

79. A person holds two basketballs. What is the net torque about A?

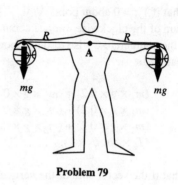

Problem 79

    A. Zero.
    B. Non-zero.

80. Suppose there is only one basketball in problem **79**. What is the net torque about point A?

    A. $m \times g \times R$ clockwise.
    B. $m \times g \times R$ counterclockwise.
    C. Zero.

For problems **81-83**, a ladder stands against a wall (assume friction is present on the wall). The ladder is 4 m long and makes an angle of 37° with the horizontal at the base.

81. What if $\sum \tau = 0$ about point A?
(Sum of clockwise torques about the axis of rotation = sum of counterclockwise torques about the same axis of rotation.)

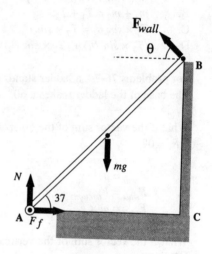

Problems 81-83

    A. $(F_{wall} \times \cos \theta \times 4 \times \sin 37) +$
          $(F_{wall} \times \sin \theta \times 4 \times \cos 37) =$
             $m \times g \times [(4 \times \cos 37) / 2]$.
    B. $F_{wall} \times \cos \theta \times 4 \times \sin 37 = [m \times g \times (4 \times \cos 37) / 2]$
        $+ (F_{wall} \times \sin \theta \times 4 \times \cos 37$.
    C. $(F_{wall} \times \cos \theta \times 4 \times \cos 37) +$
          $(F_{wall} \times \sin \theta \times 4 \times \sin 37) =$
             $m \times g \times [(4 \times \cos 37) / 2]$.
    D. $(F_{wall} \times \cos \theta \times 4) + (F_{wall} \times \sin \theta) - (m \times g) + F_f = 0$.

82. What if the vector sum of the horizontal force components $\sum F_x = 0$?

    A. $F_{wall} \times \cos \theta = F_f$.
    B. $F_{wall} \times \sin \theta = F_f$.
    C. $N = F_f$.

83. What if the vector sum of the vertical force components $\sum F_y = 0$?

    A. $(F_{wall} \times \sin \theta) + N = F_f$.
    B. $(F_{wall} \times \sin \theta) + F_f = N + (m \times g)$.
    C. $(F_{wall} \times \sin \theta) + N = m \times g$.

84. A tree branch of length R is pulled with twine as shown. What is the net torque about point A? For simplicity, assume mass is uniformly distributed for the tree branch.

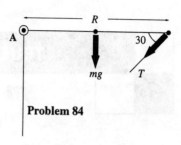

Problem 84

A. $(T \times \cos 30 \times R) + [m \times g \times (R / 2)]$ clockwise.
B. $(T \times \cos 30 \times R) + [m \times g \times (R / 2)]$ counterclockwise.
C. $(T \times \sin 30 \times R) + [m \times g \times (R / 2)]$ clockwise.
D. $T \times \sin 30 \times R$ clockwise.

For problems **85-87**, a cat of mass m_1 climbs onto a gate of mass m_2. The gate has only one hinge (to simplify the problem).

85. What if $\sum \tau = 0$ about point A?
(Sum of clockwise torques about the axis of rotation = sum of counterclockwise torques about the same axis of rotation.)

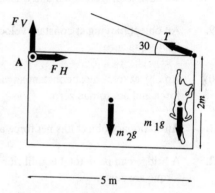

A. $T \times \sin 30 \times 2 = (m_1 \times g \times 5) + (m_2 \times g \times 2.5)$.
B. $T \times 2 = (m_1 \times g \times 5) + (m_2 \times g \times 2)$.
C. $T \times \sin 30 \times 5 = (m_2 \times g \times 2.5) + (m_1 \times g \times 5)$.
D. $T \times \cos 30 \times 5 = (m_2 \times g \times 2.5) + (m_1 \times g \times 5)$.

Problems 85-87

86. What if the vector sum of the horizontal force components is $\sum F_x = 0$?

A. $F_H = T$.
B. $F_H = T \times \cos 30$.
C. $F_H = T \times \sin 30$.

87. What if the vector sum of the vertical force components $\sum F_y = 0$?

A. $F_V + (T \times \sin 30) = (m_1 \times g) + (m_2 \times g)$.
B. $F_V + (T \times \cos 30) = (m_1 \times g) + (m_2 \times g)$.
C. $F_V = (m_1 + m_2) \times g$.

	TRUE?		CONCEPT CHECK	

m u s t K n O w

88. When the net force and the net torque on an object are zero, we have total (rotational and translational) equilibrium.

89. An object moving at constant velocity can have net force and net torque zero.

90. An object rotating at constant angular velocity can have net force and net torque zero.

91. An stationary object has net force and net torque both zero.

92. A bridge can be in total equilibrium (both net force and net torque are zero).

93. A person can be in total equilibrium (both net force and net torque are zero).

94. The earth, as it goes around the sun, is not an object that is in equilibrium.

95. One of the conditions of total equilibrium is that the (vector) sum of the horizontal components of the forces acting on an object is zero.

96. One of the conditions of total equilibrium is that the (vector) sum of the vertical components of the forces acting on an object is zero.

97. "Net torque acting on an object about an axis is zero" means that the sum of all the clockwise torques is equal to the sum of all the counterclockwise torques about that axis of rotation.

98. If the net torque about an axis is zero, then the angular acceleration of the object about that axis is zero.

99. If an object is at rest initially, then we have to apply a net torque on it to make it rotate.

100. If a body is at rest, then the net torque on the object about any axis should be zero.

101. The weight can cause an object to rotate.

102. A uniform 1-m ruler is balanced on a fulcrum. The ruler will tilt if the force of gravity does not act at the fulcrum, as the weight will then produce an unbalanced torque to tilt it one way or the other.

103. If two children of the same mass ride on a seesaw of uniformly distributed mass and sit equally far from the fulcrum, then the net torque is equal to zero. Assume they sit on opposite sides of the fulcrum.

104. If a father and a child ride on a seesaw of uniformly distributed mass and sit on opposite sides of the fulcrum which is at the middle. The father has to sit very close to the fulcrum if they are to be in equilibrium, but the child has to sit very far away.

105. In problem **104,** the weight of the father (the bigger vertical force) produces a torque about the axis of rotation in a certain direction. The child (the smaller vertical force) produces a torque in the opposite direction.

106. For problem **105,** under the action of the forces mentioned, we can say that: bigger force × smaller perpendicular distance to the axis of rotation = smaller force × larger perpendicular distance to the axis of rotation.

107. For problem **105,** the weight of the seesaw will not produce a torque as long as the fulcrum (the axis of rotation) is placed at the center of gravity.

108. For problem **105,** the perpendicular distance to the axis of rotation is zero as weight acts exactly at the fulcrum.

Rotational Motion

9

Circuit Breaker: **Warning!!!** This is a tough chapter – but you can do it! Let's go!

1. When a body goes around a circle of constant radius r with a uniform speed v, what kind of acceleration does it have?

    A. None.
    B. Centripetal.
    C. Tangential.
    D. Angular.

    For problems **2–4**, a body is moving in a circular path of radius r.

2. What is the direction of the angular acceleration α ?

    A. Along the tangent to the circle.
    B. Along the radius.
    C. Perpendicular to the plane of the circle.
    D. Along the circumference of the circle.

3. What is the magnitude of the tangential acceleration a_t?
    Assume α is the angular acceleration and r is the radius. Note that this is a very important relation!

    A. $a_T = r \times a$.
    B. $a_T = r \times \alpha$.
    C. $a_T = r \, / \, a$.
    D. $a_T = r \, / \, \alpha$.

4. Is the magnitude of the radial acceleration ($a_C = v^2 \, / \, r$) a constant?

    A. Yes.
    B. No. It is changing with time.

    For problems **5–9**, a circular disc of radius r (= 0.2 m) is rotating about a vertical axis passing through the center and perpendicular to its plane. The speed of rotation ω changes from an initial value of 20 revolutions per second to a final value of 30 revolutions per second in a time of 2 seconds.

5. What is the average angular acceleration α (in rad/s^2)?

    A. $10 \times \pi = 31.4$ rad/s^2.
    B. $20 \times \pi = 62.8$ rad/s^2.
    C. $30 \times \pi = 94.2$ rad/s^2.

6.   How many rotations [= (the angle $\Delta\theta$ in radians) / (2π radians per revolution)] does the disc make during this 2-s time interval?

      A.   40.
      B.   50.
      C.   60.
      D.   80.

7.   What is the tangential acceleration of a point on the rim, a_T (in m/s^2)?

      A. $4 \times \pi = 12.56$ m/s^2.
      B. $3 \times \pi = 9.42$ m/s^2.
      C. $2 \times \pi = 6.28$ m/s^2.
      D. $\pi = 3.14$ m/s^2.

8.   What is the angular velocity (= ω_f) of a particle on the rim at time $t = 1.5$ s (in rad/s)?

      A. $55 \times \pi = 172.7$ rad/s.
      B. $40 \times \pi = 125.6$ rad/s.
      C. $60 \times \pi = 188.4$ rad/s.
      D. $50 \times \pi = 157$ rad/s.

9.   What is the resultant total acceleration [$a = (a_T{}^2 + a_R{}^2)^{1/2}$] of a particle on the rim at time $t = 1.5$ s (in m/s^2)?
a_T is the tangential acceleration and a_R is the radial or centripetal acceleration.

      A. $55 \times \pi = 172.7$ m/s^2.
      B.   5965 m/s^2.
      C.   1900 m/s^2.
      D.   3200 m/s^2.

For problems **10–13**, a horizontal circular disc rotating about a vertical axis starts from rest accelerates at a constant rate, makes an angular displacement (angle with respect to a reference line) of 50 revolutions in 5 s, and settles to a constant angular speed at the end of 5 s.

10.   What is the average angular acceleration α of the disc (in rad/s^2)?

      A. $8 \times \pi = 25.12$ rad/s^2.
      B.   4 rad/s^2.

11.   What is the angular velocity $(= \omega_f)$ 3 s after the
       start (in rad/s)?

       A.   24 rad/s.
       B. $24 \times \pi = 75.36$ rad/s.

12.   How many rotations occur in 3 s?
       [Hint: Number of rotations =
       (the angle $\Delta\theta$ in radians) / (2π radians per rotation).]

       A.   30.
       B.   12.
       C.   18.
       D.   10.

13.   How many rotations occur in 7 s?
       [Hint: Number of rotations in 7 s = Number of rotations in the
       first 5 s + Number of rotations in the next 2 s.]

       A.   90.
       B.   40.
       C.   150.
       D.   98.

       For problems **14–15**, a body starts from rest and moves in a
       circle of radius R with uniform angular acceleration α.

14.   What is the radial (centripetal) acceleration at time t?

       A. $R \times \alpha^2 \times t^2$.
       B. $R \times \alpha \times t$.
       C. $R^2 \times \alpha^2 \times t$.
       D. $R^2 \times \alpha \times t$.

15.   What is the resultant total acceleration at time t?

       A. $R \times \alpha \times (1 + \alpha \times t^2)$.
       B. $R \times \alpha \times (1 + \alpha^2 \times t^4)^{1/2}$.
       C. $R \times \alpha \times (1 + \alpha^2 \times t)^{1/2}$.
       D. $R \times \alpha \times (1 - \alpha \times t^2)$.

For problems **16-29**, two point masses m_1 and m_2 are connected by a massless rod PQ of length l. Assume that the system's center of mass is at the origin.

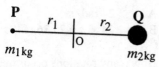

Problems 16-17

16. Masses m_1 and m_2 are placed at distances r_1 and r_2 from the axis of rotation. What is the system's moment of inertia?

    A. $I_{\text{ABOUT O}} = (m_1 \times r_1{}^2) + (m_2 \times r_2{}^2)$.
    B. $I_{\text{ABOUT O}} = (m_1 \times r_1) + (m_2 \times r_2)$.

17. In general, is the center of mass always at the axis of rotation?

    A.   Yes.
    B.   No.

18. If the bodies are not point masses, can the center of mass of a two-body system lie deep within one of the bodies? [Assume the earth-sun system.]

    A.   Yes.
    B.   No.

19. If the bodies are not point masses, can the center of mass of a two-body system lie outside the two bodies?

    A.   Yes.
    B.   No.

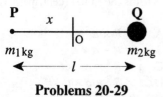

Problems 20-29

20. What is the moment of inertia of the system about an axis passing through the center of mass, with the coordinates defined as in the figure? Assume x is a positive number.

    A. $I = (m_1 \times x^2) + [m_2 \times (l - x_{\text{MAGNITUDE}})^2]$.
    B. $I = (m_1 \times x) + [m_2 \times (l - x_{\text{MAGNITUDE}})]$.

21. A system's center of mass can be determined using the formula: $x_{cm} = [(m_1 \times x_1) + (m_2 \times x_2)] / (m_1 + m_2)$. Apply the center of mass formula to the system.

    A. $x_{cm} = [m_1 \times (-x)] +$
              $[m_2 \times (l - x_{\text{MAGNITUDE}})] / (m_1 + m_2)$.
    B. $x_{cm} = [m_1 \times (+x)] +$
              $[m_2 \times (l - x_{\text{MAGNITUDE}})] / (m_1 + m_2)$.

22. If the center of mass is chosen as the origin, then:
$0 = x_{cm} =$
$\{[m_1 \times (-x)] + [m_2 \times (l - x_{\text{MAGNITUDE}})]\} / (m_1 + m_2).$

   A.   True.
   B.   False.

23. If the center of mass is chosen as the origin, then:
$0 = x_{cm} = \{[m_1 \times (-x)] + [m_2 \times (l - x_{\text{MAGNITUDE}})]\}.$

   A.   True.
   B.   False.

24. If the center of mass is chosen as the origin:
$x_{cm} = [m_2 / (m_1 + m_2)] \times l.$

   A.   True.
   B.   False.

25. If the origin is chosen as the center of mass:
$l - x_{\text{MAGNITUDE}} = l \times [m_1 / (m_1 + m_2)].$

   A.   True.
   B.   False.

26. The rotational inertia, I, of the system about an axis passing through the center of mass and perpendicular to *length l* is:
$[(m_1 \times x_{\text{MAGNITUDE}}^2) + [m_2 \times (l - x_{\text{MAGNITUDE}})^2].$

   A.   True.
   B.   False.

27. What is the moment of inertia I about an axis passing through the center of mass and perpendicular to PQ?
{Hint: $I = [(m_1 \times x_{\text{MAGNITUDE}}^2) + [m_2 \times (l - x_{\text{MAGNITUDE}})^2].$
$I = m_1 \times [(m_2 \times l) / (m_1 + m_2)]^2 +$
$\quad\quad m_2 \times [(m_1 \times l) / (m_1 + m_2)]^2.\}$

   A. $[(m_1 \times m_2) / (m_1 + m_2)] \times l^2.$
   B. $(m_1 + m_2) \times (l^2 / 4).$
   C. $[(m_1 + m_2) / 2] \times l^2.$
   D. $(m_1 \times m_2 \times l^2) / [2 \times (m_1 + m_2)].$

28. What is I about an axis passing through P and perpendicular to PQ?

    A. $m_1 \times l^2$.
    B. $m_2 \times l^2$.
    C. $(m_1 + m_2) \times l^2$.
    D. $[(m_1 \times m_2) / (m_1 + m_2)] \times l^2$.
    E.   Zero.

29. What is I_{PQ} about an axis passing through points P and Q?

    A. $m_1 \times l^2$.
    B. $m_2 \times l^2$.
    C.   Zero.
    D. $(m_1 + m_2) \times l^2$.

30. On a light rod PQ of length 5 m, masses 1, 2, 3, 4, and 5 kg are placed at distances 0, 1, 2, 3, and 4 m respectively from end P. The moment of inertia of the system about an axis through Q and perpendicular to PQ is:

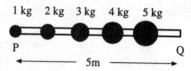

    A.   105 kg $\times$ m^2.
    B.   35 kg $\times$ m^2.
    C.   80 kg $\times$ m^2.
    D.   100 kg $\times$ m^2.

Problem 30

31. A ring of mass 2 kg and radius 0.1 m is suspended by a light massless wire of length 0.5 m from a point P. The moment of inertia of the ring about an axis through P and perpendicular to the plane of the ring is:

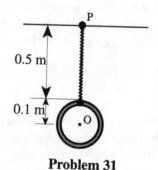

    A.   0.20 kg $\times$ m^2.
    B.   0.74 kg $\times$ m^2.
    C.   0.70 kg $\times$ m^2.
    D.   0.50 kg $\times$ m^2.

Problem 31

32. The distance between the H and Cl atoms in an HCl molecule is 1.2×10^{-11} m; the atomic weight of Cl is 35 and the mass of an H atom is 1.7×10^{-27} kg. What is the moment of inertia of the molecule about an axis through its center of mass and perpendicular to the axis of the molecule?

    A. 24×10^{-50} kg $\times$ m^2.
    B. 25×10^{-27} kg $\times$ m^2.
    C. 35×10^{-22} kg $\times$ m^2.
    D. 36×10^{-49} kg $\times$ m^2.

33. Assume that the moment of inertia of a thin uniform rod of mass m and length l about an axis through its center mass and perpendicular to its length is $[(m \times l^2)/ 12]$. What is I about an axis through the end and perpendicular to the length?

    A. $[(m \times l^2)/ 2]$ kg×m^2.
    B. $[(m \times l^2)/ 3]$ kg×m^2.
    C. $[(m \times l^2)/ 8]$ kg×m^2.
    D. $[(m \times l^2)/ 4]$ kg×m^2.

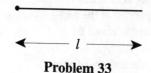

Problem 33

34. What is the moment of inertia of a solid cylinder of length 0.3 m, radius 0.05 m, and mass 10 kg about the (cylindrical) axis of the cylinder?

    A. 1.25×10^{-2} kg × m^2.
    B. 1.25 kg × m^2.
    C. 0.90 kg × m^2.
    D. 0.45 kg × m^2.

35. For problems **35–37**, PQRS is a light square frame of side 20 cm. Masses 100, 200, 300, and 400 g are placed at P, Q, R, and S, respectively. Are the center of mass of the system and the center of the square the same?

    A. Yes.
    B. No.

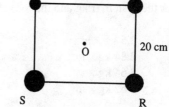

Problems 35-37

36. How far is the center of the square from each mass in cm?

    A. 14. 14 cm.
    B. 28.28 cm.
    C. 10 cm.
    D. 6.4 cm.

37. What is the moment of inertia of the system about an axis through the center O and perpendicular to the plane of the square?

    A. 2 kg × m^2.
    B. 2×10^4 kg × m^2.
    C. 2×10^5 kg × m^2.
    D. 0.0200 kg × m^2.

For problems **38–40**, a heavy rod of length 1.0 m and mass 0.6 kg is attached to a ring of mass 2 kg and radius 0.1 m. It is suspended at the end A by a horizontal nail. Find the moment of inertia about the nail (in kg × m^2).

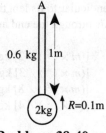

Problems 38-40

38. For the rod about point A:

    A.   0.06 kg × m^2.
    B.   0.05 kg × m^2.
    C.   0.2 kg × m^2.
    D.   0.1 kg × m^2.

39. For the ring about point A:

    A.   2.44 kg × m^2.
    B.   0.01 kg × m^2.
    C.   2.82 kg × m^2.
    D.   0.43 kg × m^2.

40. For the system (ring and the rod) about point A:

    A.   2.5 kg × m^2.
    B.   2.84 kg × m^2.
    C.   2.64 kg × m^2.
    D.   0.62 kg × m^2.

41. For problems **41–42**, a constant force F is acting at a point P. What is the torque of F about a point O distance r from P?

    A. $F \times r \times \sin \theta$.
    B. $F \times r \times \cos \theta$.
    C. $F \times r$.
    D. $F \times r \times \tan \theta$.

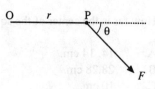

Problem 41

42. The torque of F about O is in what direction?

    A.   Clockwise.
    B.   Counterclockwise.

43. Translatory motion is produced by:

    A.   Force.
    B.   Torque.

44. Rotatory motion is produced by:

    A.   Force.

    B.   Torque.

For problems **45–51**, choose the larger moment of inertia from the two choices. The centered black circles represents the axis of rotation perpendicular to the plane of the paper. Otherwise the axis will be shown as a vertical line.

45. A circular ring of mass *m*.

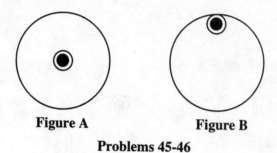

Figure A   **Figure B**

Problems 45-46

46. A sphere of mass *m*. Use the same figure as **45**.

47. A rectangular block of mass *m*.

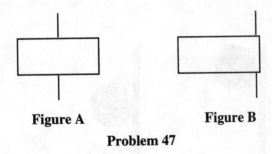

Figure A   **Figure B**

Problem 47

48. A rod of length *l*.

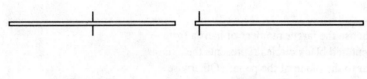

Figure A   **Figure B**

Problem 48

49. A disc of mass *m*.

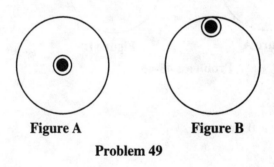

Figure A   **Figure B**

Problem 49

50. A child of mass *m* on a pole.

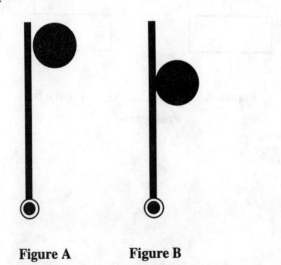

Figure A   **Figure B**

51.   A father carries a child as shown (system is father and child together).

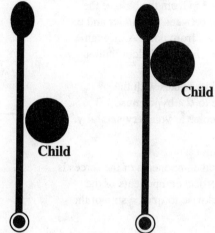

52.   For a rigid body to be in equilibrium, what is the necessary and sufficient condition? Choose the best answer.

A.   The vector sum of the forces should be zero.
B.   The vector sum of the torques about any point should be zero.
C.   The vector sum of the forces and torques about any point on the body should be zero.

53.   A meter stick PQ of weight W and center O is suspended by two spring balances at L and M. Weights W_1 and W_2 are adjusted at E and F so that the stick remains horizontal. If R_1 and R_2 are the readings of the spring balances, describe the equilibrium (translational and rotational) condition. Assume R_1 and R_2 act vertically up.

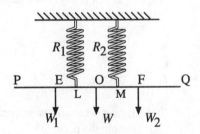

Problem 53

   [Total Equilibrium:
•   The vector sum of the horizontal components of the forces is zero.
•   The vector sum of the vertical components of the forces is zero.
•   Sum of the clockwise torques = sum of the counterclockwise torques.]

A. $[R_1 + R_2] = [W + W_1 + W_2]$ and $[(W_1 \times \text{PE}) + (W \times \text{PO}) + (W_2 \times \text{PF})] = [(R_1 \times \text{PL}) + (R_2 \times \text{PM})]$.
B. $[R_1 + R_2] = [W_1 + W_2]$.
C. $[(W_1 \times \text{PE}) + (W \times \text{PO})] = [(W_2 \times \text{PF}) + (R_1 \times \text{PL}) + (R_2 \times \text{PM})]$.

54. A uniform ladder of length 5 m and mass 20 kg rests against a smooth vertical wall with its foot 3 m from the base of the wall. If the coefficient of friction between the ladder and the ground is 0.4, what is the distance x from point A through which a person of mass 60 kg can climb the ladder without slipping?
[Hint: θ is the angle that the ladder makes with the horizontal. $cos\ \theta$ = adjacent side to θ / hypotenuse = 3 / 5. At this point these tricks should come to you very naturally. You can now solve for the angle.
Also, use total equilibrium conditions:
• The vector sum of the horizontal components of the forces is zero. • The vector sum of the vertical components of the forces is zero. • Sum of the clockwise torques = sum of the counterclockwise torques.]

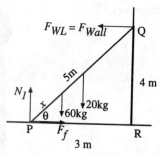

Problem 54

    A.   3.83 m.
    B.   1.17 m.
    C.   2.50 m.
    D.   2.72 m.

55. A uniform meter stick AB of weight W_M (= 2 N) is resting symmetrically on two supports C and D 0.2 m apart. What weight W_B (this is shown as W in the figure) should we suspend at end B to just tilt the stick away from C?

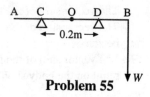

Problem 55

    A.   2 N.
    B.   4 N.
    C.   0.5 N.
    D.   0.67 N.

56. In the figure, what are the respective torques of the forces F_1 and F_2 about the point O?

    A. $F_1 \times d_1$ and $F_2 \times d_2$ in opposite directions.
    B. $F_1 \times d_1$ and $F_2 \times d_2$ in the same direction.
    C. $F_1 \times d_2$ and $F_2 \times d_1$ in opposite directions.
    D. $F_1 \times d_2$ and $F_2 \times d_1$ in the same direction.

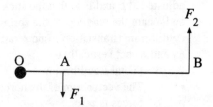

Distance OA = d_1

Distance OB = d_2

Problem 56

57. A uniform meter stick AB placed on a support C is balanced when a mass of 0.5 kg is placed at A. If AC = 0.3 m, what is the *mass* of the stick (in kg)? Assume the weight of the meter stick W acts in the middle (as shown in the figure) at point O.

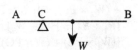

Problem 57

    A.   0.5 kg.
    B.   0.75 kg.
    C.   1.0 kg.

58. A non-uniform stick AB of length 1.0 m balances on a support at G where AG = 0.6 m. If the support is shifted to the center C of the stick, it balances only when a mass of 0.2 kg is placed at A. What is the mass of the stick (in kg)?
[Hint: When the support is moved to point C, weight of the stick at point G now causes a torque that is clockwise.]

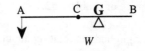

Problem 58

   A. 1.0 kg.
   B. 0.2 kg.
   C. 0.5 kg.
   D. 1.2 kg.

59. Equal and opposite forces of magnitude F act at points A and B. What is the net torque about the points C, D, and E, respectively?

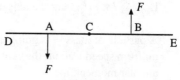

Problem 59

   A. $F \times AC$, $F \times AD$, and $F \times AE$.
   B. All are $F \times AB$ counterclockwise.
   C. $F \times BC$, $F \times BD$, and $F \times BE$.
   D. $F \times AC$, $F \times BD$, and $F \times BE$.

60. Two equal and opposite forces F act at points A and B of a rigid body, and d is the perpendicular distance between the lines of action of the forces. What are the resultant force on the body and the resultant torque about a point A on the body?

   A. Zero and $(F \times d)$.
   B. Zero and $(F \times AB)$.
   C. $2 \times F$ and $(F \times d)$.
   D. $2 \times F$ and $(F \times AB)$.

Problem 60

61. When equal and opposite forces act at two different points on a free rigid body (please look at the figure), what motion results?

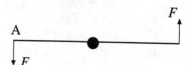

   A. No translatory or rotatory motions.
   B. Translatory motion of its center of mass but no rotatory motion.
   C. Rotatory motion but no translatory motion of its center of mass.

Problem 61

62. Are the center of mass of a rigid body and its center of gravity at the same point on the surface of the earth? (Assume earth is a uniform sphere.)

   A. Yes.
   B. No.

63.   Does the earth-moon system have a center of mass? Use the choice of problem **62**.

64.   A rigid body is rotating about an axis with uniform angular velocity ω and moment of inertia I about the axis. What is its angular momentum?

    A. $I \times \omega$ perpendicular to the axis of rotation.
    B. $I \times \omega$ along the axis of rotation.
    C. $I \times \omega^2$ perpendicular to the axis of rotation.
    D. $I \times \omega^2$ along the axis of rotation.

65.   A particle of mass m is going around a circle of radius r with uniform speed v about the center of the circle. What is its angular momentum?

    A. $[m \times v \times r]$ along the tangent to the circle.
    B. $[m \times v]$ along the tangent to the circle.
    C. $[m \times v \times r]$ along the line perpendicular to the plane of the circle.
    D. $[(m \times v^2) / r]$ along a radius.

For problems **66–67**, a circular disc of mass m and radius $r\,(= R / 2)$ is placed with its center at $R / 2$ <u>on top</u> of another disc of radius R and mass M. Distance $AB =\ R / 2$.

66.   The system rotates about an axis through A and perpendicular to the plane of the disc. What is its moment of inertia?

    A. $\{M - [(3/4) \times m)]\} \times R^2 / 2.$
    B. $(M + m) \times R^2 / 2.$
    C. $(M - m) \times R^2 / 2.$
    D. $\{M + [(3/4) \times m]\} \times R^2 / 2.$

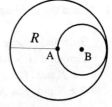

Problems 66-67

67.   The system rotates about an axis through B and perpendicular to the plane of the discs. What is the moment of inertia?

    A. $\{[(3/4) \times M] + (m / 8)\} \times R^2.$
    B. $\{[(3/4) \times M] - (m / 8)]\} \times R^2.$
    C. $(M + m) \times R^2.$
    D. $(M - m) \times R^2.$

For problems **68-69**, a circular disc has mass m, radius R, and center at A. A circular portion of radius $R/2$ is now *removed* from A. Find the moment of inertia of the remaining portion for the situations described.

68. What is the moment of inertia through A and perpendicular to the plane of the disc? Use the same figure as **67**.

    A. $(3/8) \times m \times R^2$.
    B. $(13/32) \times m \times R^2$.
    C. $(23/32) \times m \times R^2$.
    D. $(m \times R^2)/4$.

69. What is the moment of inertia through B and perpendicular to the plane of the disc?

    A. $(23/32) \times m \times R^2$.
    B. $(13/32) \times m \times R^2$.
    C. $(m \times R^2)/7$.
    D. $(m \times R^2)/3$.

For problems **70-75**, a solid cylinder of mass m_1 and radius R can rotate freely about its horizontal (cylindrical) axis. A mass $4 \times m_1$ is held at a height h above the ground by being attached to a light thin string wound around the cylinder. When released, mass m_2 accelerates downward with an acceleration a.

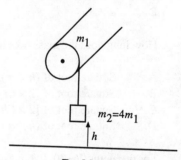

Problems 70-75

70. When the mass $4 \times m_1$ is just released from rest, what can we say about tension acting on mass m_2? Assume that the system accelerates beginning immediately.

    A.   Tension is more than the weight $4 \times m_1 g$.
    B.   Tension is equal to the weight $4 \times m_1 g$.
    C.   Tension is less than the weight $4 \times m_1 g$.

71. When the mass $m_2 (= 4 \times m_1)$ is released, what is its acceleration a ?

    A. $(8/9) \times g$ $= 8.9$ m/s^2.
    B. g $= 10$ m/s^2.
    C. $4 \times g$ $= 40$ m/s^2.

72. What is the tension T of the string when the mass $4 \times m_1$ accelerates?

    A. $m_1 \times a$.
    B. $(m_1 \times a) / 2$.
    C. $(m_1 \times a) / 9$.
    D. $5 \times m_1 \times a$.

73. What is the net torque τ_{NET} about the axis of the cylinder when a mass $m_2 (= 4 \times m_1)$ accelerates?

    A. $4 \times m_1 \times g \times R$.
    B. $(m_1 \times g \times R) / a$.
    C. $(5 \times m_1 \times g \times R) / a$.
    D. $(m_1 \times R \times a) / 2$.

74. What is the angular acceleration α of the cylinder?

    A. R / a.
    B. $a \times R$.
    C. $a / (2 \times R)$.
    D. a / R.

75. How long does the mass take to touch the floor?

    A.   Square root of (h / g).
    B.   Square root of $[2 \times (h / g)]$.
    C.   Square root of $[2 \times (h / a)]$.
    D.   Square root of (h / a).

For problems **76–80**, two masses of 4 and 20 kg are connected by a string passing over a pulley, which is a cylindrical disc of mass m and radius 0.1 m rotating about its horizontal axis. The 20-kg mass is held at a height of 2 m above the ground and then released. It takes 0.8 s to reach the ground at constant acceleration. The pulley rotates such that the string never slips.

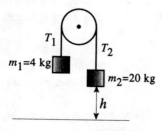

Problems 76-80

76. What is the acceleration of the system (in m/s^2)?

    A.   6.25 m/s^2.
    B.   2.5 m/s^2.
    C.   12.5 m/s^2.
    D.   10 m/s^2.

77. What is the angular acceleration of the pulley (in rad/s^2)?

    A.   125 rad/s^2.
    B.   12.5 rad/s^2.
    C.   62.5 rad/s^2.
    D.   31.25 rad/s^2.

78. What is the tension in the two parts of the string (in N)?

    A. $T_1 = 75$ N, and $T_2 = 65$ N.
    B. $T_1 = 65$ N, and $T_2 = 75$.
    C. $T_1 = 65$ N, and $T_2 = 65$.

79. What is the moment of inertia of the pulley about an axis passing through its center?
(Assume that the string does not slip on the pulley.)

    A.   125 kg × m^2.
    B.   0.8 kg × m^2.
    C.   65 kg × m^2.
    D.   0.016 kg × m^2.

80. What is the angular velocity of the pulley as the mass touches the ground (in rad/s)?

    A.   25 rad/s.
    B.   50 rad/s.
    C.   12.5 rad/s.
    D.   100 rad/s.

81. What is the rotational counterpart of mass *m* in linear motion?

    A.   Moment of inertia.
    B.   Torque.
    C.   Angular momentum.

82. What is the rotational counterpart of force in linear motion?

    A.   Moment of inertia.
    B.   Torque.
    C.   Angular momentum.

83. What is the rotational counterpart of momentum in linear motion?

    A.   Moment of inertia.
    B.   Torque.
    C.   Angular momentum.

84. Two masses M and m are connected by a string passing over a pulley. If the pulley is heavy, the acceleration of the system is a_1 and if the pulley is light and of the same radius R, the acceleration is a_2. Assume no slipping between the string and the pulley. Compare the accelerations.

    A. $a_1 = a_2$.
    B. $a_1 > a_2$.
    C. $a_1 < a_2$.

Problem 84

85. If the net torque is zero then what statement is always true?

    A.   Net force is zero.
    B.   Angular momentum is conserved.
    C.   Linear momentum is conserved.
    D.   Angular momentum is zero.

86. Can an object rotate through 24,563 degrees?

    A.   Yes.
    B.   No.

87. What is 1 radian of apple pie?

    A.   87.3 degrees.
    B.   57.3 degrees.

For problems **88–89**, a mass m is suspended by a string passing over a pulley which (supported by an axle through the center) is a circular disc of radius R and mass M.

88. What forces are acting on the pulley?

    A. T, $(M \times g)$, and reaction N of the axle.
    B. T.
    C. T and $(m \times g)$.
    D. $(m \times g)$ and R.

Problems 88-89

89. What forces contribute to torque about the axle of the pulley?

    A. $(m \times g)$ and R.
    B. T.
    C. $(m \times g)$.
    D. R.

For problems **90–92**, a spherical body has moment of inertia I about its own axis (which is perpendicular to its direction of motion and parallel to the surface of the plane). The body is rolling down an incline of angle θ without slipping.

90. If the sphere has radius R, what is its acceleration?
[Hint: Use energy methods and equations of kinematics.]

    A. $(g \times sin\ \theta)$.
    B. g.
    C. $7 \times sin\ \theta$.
    D. $(g \times sin\ \theta)\ /\ 2$.

91. If the body is a solid cylinder of radius R, what is its acceleration?

    A. g.
    B. $(g \times sin\ \theta)$.
    C. $(5/\ 7) \times g \times sin\ \theta$.
    D. $(2/\ 3) \times g \times sin\ \theta$.

92. If the body is a ring of radius R, what is its acceleration?

    A. $(g \times sin\theta)$.
    B. g.
    C. $(g \times sin\ \theta)\ /\ 2$.
    D. $(2/\ 3) \times g \times sin\ \theta$.

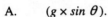

Problems 90-92

For problems **93–96**, a flywheel of radius R and mass m is rotating about its axis with angular velocity ω. The flywheel is a solid uniform disk.

93. What is the kinetic energy due to rotation?

    A. $[(m \times R^2 \times \omega^2)\ /\ 4]$.
    B. $[½ \times (m \times R^2 \times \omega^2)]$.
    C. $[m \times R \times \omega^2]$.
    D. $[(m \times R^2 \times \omega^2)\ /\ 4]$.

94.   What is the average torque to stop it in time t?

    A. $(m \times R^2 \times \omega_0) / t$.
    B. $(m \times R^2 \times \omega_0) / (2 \times t)$.
    C. $m \times R^2 \times \omega_0 \times t$.
    D. $(m \times R^2 \times \omega_0 \times t) / 2$.

95.   What is the average torque to stop it in n rotations?

    A. $(m \times R^2 \times \omega^2) / (\pi \times n)$.
    B. $(m \times R^2 \times \omega^2) / (2 \times \pi \times n)$.
    C. $(m \times R^2 \times \omega^2) / (8 \times \pi \times n)$.
    D. $(m \times R \times \omega^2) / (\pi \times n)$.

96.   What is the average force applied on the rim to stop it in time t?

    A. $(m \times R \times \omega) / t$.
    B. $(m \times R^2 \times \omega) / t$.
    C. $(m \times R \times \omega) / (4 \times t)$.
    D. $(m \times R \times \omega) / (2 \times t)$.

For problems **97–102**, a cylinder of radius R and mass m rolls down an incline of inclination θ to the horizontal, through a distance l down the plane.

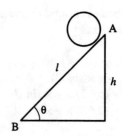

Problems 97-102

97.   What is the speed at B?

    A.   Square root of $[(4 \times g \times l \times sin\ \theta) / 3]$.
    B.   Square root of $[(g \times l \times sin\ \theta) / 3]$.
    C.   Square root of $[(g \times l \times sin\ \theta) / 6]$.
    D.   Square root of $(2 \times g \times l)$.

98.   What is the kinetic energy at the base? Assume no friction.

    A. $(m \times g \times l)$.
    B. $(m \times g \times l \times sin\ \theta)$.
    C. $(m \times g \times l \times sin\ \theta) / 2$.
    D. $(m \times g \times l) / 2$.

99.   What is the time taken to travel l meters?

    A. $l / (g \times sin\ \theta)$.
    B.   Square root of $[l / (g \times sin\ \theta)]$.
    C.   Square root of $[(3 \times l) / (g \times sin\ \theta)]$.
    D.   Square root of $[(4 \times l) / (g \times sin\ \theta)]$.

100. What is the force acting downward parallel to the inclined plane at its center of mass?

    A. $m \times g \times \cos \theta$.
    B. $m \times g \times \sin \theta$.
    C. $m \times g$.
    D. $(m \times g) / 2$.

101. What is the angular momentum of the cylinder at the base of the inclined plane?

    A. $2 \times m \times R \times$ square root of $[(g \times l \times \sin \theta) / 3]$.
    B. $m \times R \times$ square root of $[(g \times l \times \sin \theta)/ 3]$.
    C. $[(m \times R) / 2] \times$ square root of $[(4 \times g \times l \times \sin \theta) / 3]$.
    D. $m \times R \times$ square root of $(2 \times g \times l \times \sin \theta)$.

102. Is the angular momentum of the cylinder conserved during motion?

    A.   Yes.
    B.   No.

For problems **103–107**, a cylinder and a sphere of the same mass and radius roll down an inclined plane of angle θ with respect to the horizontal.

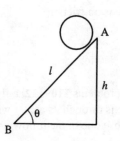

103. What is the ratio of their accelerations ($a_{\text{CYLINDER}} / a_{\text{SPHERE}}$) at the base of the inclined plane?
[Psych it out: $I_{\text{cylinder}} > I_{\text{sphere}}$.]

    A.   3.12.
    B.   0.93.

104. What is the ratio of their speeds ($v_{\text{CYLINDER}} / v_{\text{SPHERE}}$) at the base of the inclined plane?

    A.   2.4.
    B.   0.97.

105. What is the ratio of their total kinetic energies ($KE_{\text{CYLINDER}} / KE_{\text{SPHERE}}$) at the base of the inclined plane?

    A.   4.
    B.   3.
    C.   2.
    D.   1.

106. What is the ratio of their potential energies
(PE_{CYLINDER} / PE_{SPHERE}) at the start at a height h?

    A.   3.
    B.   2.
    C.   1.

107. What is the ratio of their angular momentum
(L_{CYLINDER} / L_{SPHERE}) at the base of the inclined plane?

    A.   1.9.
    B.   1.2.

108. When a body rolls down an inclined plane without slipping,
what force provides the net torque?

    A.   Gravitational force.
    B.   Frictional force.
    C.   Centripetal force.

109. Can an object roll down an inclined plane without slipping if
there is no friction?

    A.   Yes.
    B.   No.

For problems **110–112**, a fly-wheel (thick disc) rotates about
an axis through its center with constant angular velocity.
Choose one of the following choices:

    A.   Yes.
    B.   No.

110. Is the linear momentum changing for a point on the rim? Note
that the linear momentum of the center of mass is *not*
changing.

111. Is the angular momentum changing?

112. Is there a net torque on the system?

113. For the system drawn, the pulley is a disc of mass m_1 and radius R. The mass m_2 ($= 2m_1$) is placed on the inclined plane. It is connected to another mass m_3 ($= 4m_1$) by means of a string. The mass m_2 slides upwards along the plane. What is the acceleration of the system if there is no friction between mass m_2 and the surface of the inclined plane and there is no slipping of the string on the pulley?

   A. 4.6 m/s^2.
   B. 7.3 m/s^2.

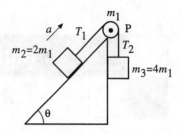

Problem 113

114. For the system drawn, the pulley is a disc of mass m_3 and radius R. The mass m_1 ($= 6m_3$) is placed on the inclined plane. It is connected to another mass m_2 ($= 2m_3$) by means of a string. The mass m_1 slides down along the plane. What is the acceleration of the system if there is no friction between mass m_1 and the surface of the inclined plane?

   A. 4.8 m/s^2.
   B. 3.2 m/s^2.
   C. 2.5 m/s^2.
   D. 1.2 m/s^2.

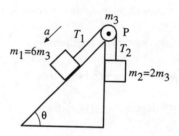

Problem 114

115. An object of weight W is lifted off the ground (barely) using the 6-pulley system. What pulling force P is required to barely lift the mass off the ground?

   A. $W / 3$.
   B. $W / 6$.

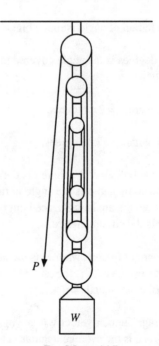

Problem 115

```
┌────────┬─────────────────────┬─────────────────────┬────────┐
│        │      TRUE?          │   CONCEPT CHECK      │        │
│        ├─────────────────────┴─────────────────────┤        │
│        │        m u s t   K   n O w                │        │
└────────┴───────────────────────────────────────────┴────────┘
```

116.   Arc length $l = R\theta$, where l is the arc length in m, R is the radius in m, and θ is the angle in radians.

117.   180 degrees = π radians.

118.   180 degrees ≈ 3.14 radians.

119.   18 degrees ≈ 0.314 radians.

120.   36 degrees ≈ 0.628 radians.

121.   A one degree angle is not the same as a one radian angle.

122.   One radian of angle is much bigger than one degree of angle.

123.   360 degrees is the angle covered in one (full or complete) rotation.

124.   360 degrees = 2π radians.

125.   One radian ≈ 57.3 degrees.

126.   If arc length $l = R\theta$, where l is the arc length in m, R is the radius in m, and θ is the angle in radians, then for one full rotation, the arc length traced out by a point at a distance R equals the circumference.

127.   Arc length $l = R\theta$, where l is the arc length in m, R is the radius in m, and θ is the angle in radians. For one full rotation, arc length = $R \times (2\pi)$.

128.   Average angular velocity = $\omega_{AVERAGE} = \Delta\theta / \Delta t$, where $\omega_{AVERAGE}$ is the average angular velocity in radians/s, $\Delta\theta$ is the angular displacement in radians, and Δt is the time interval in s.

129.   If an object's average angular velocity is a constant, then the object rotates through the same angle per unit time. Assume "unit time" is equal to the time over which we average.

130.   A fan that has reached its maximum rotational speed is an example of constant angular velocity.

131.   An airplane engine that has reached its maximum rotational speed is an example of constant angular velocity.

132.   If you drive at constant velocity, the shaft under your car rotates at constant angular velocity.

133.   The unit of angular velocity is radians per second, as the angular displacement $\Delta\theta$ is in radians and the time interval is in s.

134.   The angle covered per unit time by a rotating object can be greater than 360 degrees.

135.   1 radian = 360 degrees $/\ 2\pi$.

136.   1 degree = $(2\pi\ /\ 360)$ radians.

137.   An angular change (change of angle) of 1 radian is nearly 57 times greater than a change of 1 degree.

138.   If a shaft rotates at a constant angular velocity of 2 radians per second, then *all* points of the shaft rotate at 2 radians per second.

139.   For a rigid body, *every point* in it moves through the same angle $\Delta\theta$ in the same time interval Δt.

140.   If angular velocity is a constant, then angular acceleration is zero.

141.   Angular acceleration is the rate of change of angular velocity.

142.   If an object's angular velocity and angular acceleration have the same sign, its rotational speed increases, so it covers more angle per unit time as time progresses.

143.   If an object's angular velocity and angular acceleration have opposite signs, its rotational speed decreases, so it covers less angle per unit time as time progresses.

144. If a needle takes 60 s to complete one rotation, its angular velocity is 1 revolution per 60 seconds = 360 degrees / 60 s = 2π radians / 60 s.

145. If the earth takes 24 hr to complete one rotation about its axis, its angular velocity is 1 revolution per 24 hr = 2π radians / 24 hr = 2π radians / (24 × 3600 s) = 360 degrees / 24 hr.

146. If the earth takes 365 days to complete one revolution about the sun, its angular velocity is 1 revolution per year = 2π radians / 1 yr = 2π radians / (365 × 24 × 3600 s) = 360 degrees / (365 × 24 hr).

147. If a needle takes 12 hr to complete one rotation, its angular velocity is 1 revolution per 12 hr = 2π radians / 12 hr = 2π radians / (12 × 3600 s) = 360 degrees / 12 hr.

148. The number of rotations in 867 degrees = 867 / 360, since the angle covered in one rotation is 360 degrees.

149. The number of rotations in 867 radians = 867 / (2π), since the angle covered in one rotation is 2π radians.

150. If an airplane is at rest on the ground, at the instant you start its engine , the blades begin picking up rotational speed. Here angular acceleration and angular velocity have the same sign, hence the blades are accelerating (the rotational speed increases).

151. If an airplane is at rest on the ground, at the instant you switch off its engine, the blades begin losing the rotational speed. Here angular acceleration and angular velocity have opposite signs, hence the blades are decelerating (the rotational speed decreases).

152. When you turn the door knob or a key to open a door, there is angular displacement.

153. When a door is opened, there is angular displacement.

154. When a steering wheel is turned, there is angular displacement.

155. When you open the tap, there is angular displacement.

156. When the wheels of a car rotate, there is angular displacement.

157. When a fan rotates, there is angular displacement.

158. Imagine a circular coliseum. The number of seats that can be put in the uppermost level is always greater than what fits on the lower level, assuming a fixed distance between seats.

159. Imagine a church where seating is arranged in concentric circles. The number of people that can be seated at the back in one row is always greater than how many can be accommodated in the front row.

160. Anything that is rotating has a certain non-zero angular speed.

161. The blades of a ceiling fan rotate at a constant angular velocity once the fan reaches its maximum speed.

162. Angular velocity can be measured in radians per second.

163. Angular velocity can be expressed in rotations per second.

164. A constant angular velocity of 2,000 revolutions per minute (RPM) is the same as $(2000 \times 2\pi) / 60$ radians per second.

165. For a rotating blade inside a food processor, the linear velocities of different particles at the same distance from the axis of rotation point in all different directions.

166. The axis of rotation combined with the right-hand rule gives the direction of angular velocity.

167. The right-hand rule says that when fingers of the right hand are curled in the direction of rotation, the thumb gives the direction of angular velocity.

168. For the two bicycle wheels, the angular velocities point in the same direction if the handlebars hold the front wheel straight.

169. Angular velocity can point along $+y$.

170. Angular velocity can point along $-y$.

171. Angular velocity can point along $+x$.

172. An object rotates with a constant angular velocity of magnitude ω pointing along +y. After a few seconds, if the same object rotates with a constant angular velocity of magnitude ω pointing along -y, then the change in angular velocity is 2 ω.

173. If the angular velocity changes in magnitude (slows down or speeds up), or in direction, then we have angular acceleration α.

174. A unit of angular acceleration can be radians per second *per second*.

175. Angular acceleration is measured in rad/s^2.

176. The bigger the magnitude of angular velocity, the faster the rotational rate.

177. The bigger the angular velocity, the bigger the angle covered per unit time.

178. While rotating, if two objects rotate through the same angle in the same time interval, then they have the same average angular speed ω.

179. The angular acceleration of a rigid rotating body is the same for all points.

180. The relation for linear speed $v = R\omega$ is valid only if the angular velocity ω is in radians per second. Assume R is the radius in m.

181. The relation for linear speed $v = R\omega$ tells us that the linear speed for particles on a rotating object will be different for particles at different radii.

182. For a rotating blade of a ceiling fan, all parts of the blade have the same angular velocity.

183. For the above problem, the linear speed will be different for different radii.

184. For problem **182**, a point that is farther from the motor covers a larger circular arc length in each time interval than a point closer to the motor.

185. For problem **182**, the linear speed is higher for a point farther from the motor because it covers more arc length in unit time.

186. If you are at rest at the pole of the earth, your linear speed will be zero even though the earth is rotating once every 24 hours about its axis.

187. For problem **186** if you are at rest at the equator, your linear speed is non-zero.

188. In a 400-m dash at the Olympics, we see starting blocks at the outer track compensated in position at the start because outer lanes have a greater length than the inner lanes.

189. For any circle, circumference / radius = 2π.

190. Circumference / diameter = π.

191. The tangential linear acceleration a_T of a particle in a rotating rigid body at a distance R from the axis of rotation $a_T = R\alpha$, where α is the angular acceleration in rad/s^2.

192. A gymnast's legs can show angular displacement.

193. When a hanging traffic light swings back and forth in the wind, there is angular displacement.

194. When a traffic light swings back and forth in the wind, both tangential acceleration and angular acceleration are present. (Ignore the position of minimum height.)

195. A wheel rotating at constant angular velocity has angular displacement.

196. For problem **195**, the tangential acceleration is zero.

197. For problem **195**, the centripetal acceleration is non-zero.

198. For a roller coaster moving in vertical circles, there are tangential and angular accelerations. (Ignore the position of minimum/maximum height.)

199. The tangential acceleration a_T and angular acceleration α of a particle in a rotating body are related by $a_T = R\alpha$, where α is the angular acceleration in rad/s^2.

CHAPTER**nine**

200. The centripetal acceleration for particles moving with a speed v in a circle of radius R is given by $a_C = v^2/R$.

201. The centripetal acceleration for particles moving with a speed v in a circle of radius R is given by $a_C = \omega^2 R$, where ω is the angular speed.

202. For particles at a distance R from the center of a rotating circle, if the centripetal acceleration is constant in magnitude, then the linear speed v is constant.

203. For particles at a distance R from the center of a rotating circle, if the centripetal acceleration is constant in magnitude, then the angular speed ω is a constant.

204. Angular velocity is given by $\omega = 2\pi f$, where f is the frequency in hertz.

205. 1 hertz (the unit of frequency) $= 1/s$.

206. Average angular velocity is given by $\omega = \Delta\theta/\Delta t$, where $\Delta\theta$ is the angular displacement in radians and Δt is the time interval in s.

207. If an object is turned through 360 degrees, then the average angular velocity $\omega = 2\pi/T$, where T is the time interval for one full rotation (called its period) in s.

208. The kinematic equations for constant angular acceleration rotational motion are valid only if the angular acceleration is a constant.

209. The equation $\omega_f = \omega_0 + \alpha t$ is the same as $\alpha = (\omega_f - \omega_0)/t$.

210. If the equation $\omega_f = \omega_0 + \alpha t$ takes the form $3 = 2 + \alpha t$, the initial angular velocity is in the same direction as the angular acceleration.

211. If the equation $\omega_f = \omega_0 + \alpha t$ takes the form $2 = 3 + \alpha t$, the initial angular velocity is in the direction opposite to the angular acceleration.

212. If the equation $\omega_f = \omega_0 + \alpha t$ takes the form $-3 = 2 + \alpha t$, the initial angular velocity is in the direction opposite to the angular acceleration.

213. If the equation $\omega_f = \omega_0 + \alpha t$ takes the form $3 = (-2) + \alpha t$, the initial angular velocity is in the same direction as the angular acceleration.

214. Stephanie carried (uphill?) by Jim has more mass distributed away from the axis compared to a small child in the same position.

215. If a girl pulls in her arms as she spins about a vertical axis, her moment of inertia decreases.

216. Angular momentum points in the direction of angular velocity.

217. The earth has angular momentum about its axis of rotation.

218. The earth as it goes around the sun, has an angular momentum.

219. An electron with two units of angular momentum is rotating much faster than an electron having one unit.

220. If you spin with two empty glasses on an open air platform, the total moment of inertia will be the sum of moment of inertia of the glasses, the person, and the platform.

221. If you hold and spin two empty glasses on an open air platform, you will spin even slower if it rains and water accumulates inside the glass.

222. If a platform with a person at the center spins, both the platform and the person have the same angular velocity.

223. If a plate and doughnut have identical masses but different radii (plate has the larger radius), and if the doughnut is placed at the plate's center and both are spun, both do not have the same moment of inertia I.

224. If a plate with a doughnut at the center spins in outer space, it will continue to do so in the absence of external torques.

225. If a plate with a doughnut at the center spins, the plate will have larger moment of inertia I than the doughnut. Refer to problem **223**.

226. If a plate with a doughnut at the center spin together, the plate will have larger angular momentum L than the doughnut. Refer to problem **223**.

227. If angular momentum of a rotating object changes, there is a net torque.

228. Angular momentum is a vector because angular velocity is a vector.

229. In the absence of a net torque, angular momentum is a conserved quantity.

230. A rotating object has rotational kinetic energy if it has mass.

231. The bigger the moment of inertia, the bigger the kinetic energy for a given rotational velocity.

232. The bigger the angular speed of a body, the bigger its kinetic energy (assuming the same moment of inertia).

233. Both translational kinetic energy and rotational kinetic energy are measured in joules.

234. Tangential acceleration a_T and centripetal acceleration a_C are always at right angles.

For problems **235-254**, a massless vertical wheel of radius R with its axis of rotation at the center is held stationary in the vertical plane. It is free to rotate about its center without friction. Assume the wheel is hung from the ceilings by a cable, and that it is barely touching the ground.

Bubble gum is now stuck on the rim on the *right side*. The bubble gum is at a height $h = R$ from the ground. Immediately, the wheel begins to rotate.

235. The total energy at the start is all potential energy ($= mgR$), since the kinetic energy is zero at the start.

236. At the start, there is a tangential force on the wheel.

237. At the start, there is a net torque on the wheel due to the weight of the bubble gum acting tangentially.

238. The moment of inertia of the bubble gum about the axis of rotation is mR^2.

239. The total kinetic energy of the system is the rotational kinetic energy.

240. The rotational kinetic energy of the system at the bottom (wheel and bubble gum) is $\frac{1}{2}I\,\omega^2$, where I is the moment of inertia and ω is the angular speed at the bottom.

241. The rotational kinetic energy of the system at the bottom (wheel and bubble gum) is $\frac{1}{2}(mR^2)\,\omega^2$, where ω is the angular speed at the bottom.

242. The total energy at the bottom is all kinetic, as the potential energy is zero.

243. If we equate total energy at the start = total energy at the bottom, then we have:
$mgR = \frac{1}{2}(mR^2)\,\omega^2$.

244. This system can oscillate back and forth due to energy conservation.

245. The angular momentum $L = mvR$ changes because linear speed v is continuously changing.

246. The magnitude of the angular momentum is maximum at the bottom.

247. The tangential acceleration at the start has a magnitude equal to the magnitude of the acceleration due to gravity g.

248. The tangential acceleration at the bottom is zero as there is no net force in the tangential direction.

249. At the bottom, there is centripetal acceleration.

250. At points in between the minimum and maximum heights, the bubble gum has tangential acceleration due to the tangential component of the weight.

251. When the bubble gum is on the way down, tangential acceleration and displacement are in the same direction, so it speeds up.

252. When the bubble gum is on the way up, tangential acceleration and displacement are in the opposite directions, so it slows down.

253. At the start, there is a net torque due to the weight of the bubble gum.

254. At the bottom position, there is no net torque on the bubble gum from the weight of the bubble gum.

255. Torque is force times perpendicular distance.

256. The angular momentum for a body rotating about a fixed axis is $L = I\omega$, where I is the moment of inertia in kg-m^2 and ω is the angular velocity in radians per second.

257. If angular momentum is conserved for a rotating system, and if moment of inertia changes, then:

$$I_{BIG}\,\omega_{SMALL} = I_{SMALL}\,\omega_{BIG}.$$

258. A spinning ice skater can increase her moment of inertia by fully extending her arms, thus distributing her mass away from the axis of rotation.

259. A spinning ice skater can decrease his moment of inertia by pulling his extended arms in, thus distributing his mass towards the axis of rotation.

260. *If* the earth increases in volume without addition of mass, more particles will be distributed away from the axis of rotation. This will increase the moment of inertia.

261. If the earth increases in volume, more particles will be distributed away from the axis of rotation. If the angular momentum is conserved, the equation

$$I_{BIG}\,\omega_{SMALL} = I_{SMALL}\,\omega_{BIG}$$ tells us that the angular speed of the earth will decrease.

262. A gondola swing has two seats facing each other. A girl sits in one seat and swings to and fro. If her brother jumps on at the instant it reaches the bottom, the moment of inertia of the system is increased but the rotational speed decreases.

263. An oscillating pendulum consisting of a ball on a string is pulled from under a table through a hole in the table. As you pull the string vertically up, the moment of inertia of the pendulum decreases.

264. A gymnast is oscillating back and forth while hanging from a bar by her legs. If she catches a ball, while oscillating, she will reduce her rotational speed. Assume little momentum is delivered to the gymnast by the ball.

For problems **1-31**, a particle of mass m_{MASS} is executing simple harmonic motion about a point O between points A and B. At a certain instant in time, the particle is at point P at a distance $x_{DISPLACEMENT}$ from the equilibrium point O. Assume the spring constant is $k_{spring\ constant}$ and that the particle is moving towards point B. Ignore gravitational force. There is no frictional energy loss.

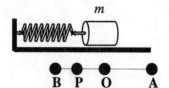

Distance OP = x

Distance OA = A

Distance OB = A
Distance AB = $2A$
Amplitude = A
Maximum Displacement = A

Problems 1-31

1. What is the magnitude of the acceleration of the particle at point P?

    A. $[k \times x] / m.$
    B. $[k \times x].$
    C. $[k \times x^2] / m.$
    D. $[k \times x^2] / m.$

2. What is the amplitude of the motion $A_{AMPLITUDE}$?

    A.   Distance AB.
    B.   Distance AP.
    C.   Distance OA = OB.
    D.   Distance BP.

3. At what point is the magnitude of acceleration of the particle a maximum?

    A.   Point O.
    B.   Point B.
    C.   Point P.
    D.   Half-way point OB/ 2.

4. Where is the magnitude of the velocity of the particle a maximum?

    A.   At point A.
    B.   At point B.
    C.   At point P.
    D.   At point O.

5. What is the period of oscillation?

    A. $k.$
    B. $1 / k.$
    C. $2\pi \times (m / k)^{1/2}.$
    D. $(2\pi)/k.$

6.   Period of oscillation is the time taken for the particle to move from:

   A.   A to O.
   B.   A to B and back to A.

For problems **7-31**, choose one of the following from the two sets of answers (whichever one is appropriate).

   A.   Yes.   A.   Zero.
   B.   No.   B.   Non-zero.

7.   What is the speed of the particle at point A?
       [Please use the second set on the right.]

8.   What is the speed of the particle at point B?
       [Please use the second set on the right.]

9.   What is the speed of the particle at point O?
       [Please use the second set on the right.]

10.   What is the kinetic energy of the particle at point A?
       [Please use the second set on the right.]

11.   What is the kinetic energy of the particle at point O?
       [Please use the second set on the right.]

12.   Does the time of oscillation depend upon the amplitude?
       [Please use the *first* set.]

13.   Does the time of oscillation depend upon the mass?
       [Please use the *first* set.]

14.   Does the time of oscillation depend upon the spring constant?
       [Please use the *first* set.]

15.   If we stretch the spring within "elastic" limits, does the spring constant change?
       [Please use the *first* set.]

16.   What is the potential energy of the system at point A?
       [Please use the second set on the right.]

17.   What is the potential energy of the system at point B?
       [Please use the second set on the right.]

18. What is the potential energy of the system at point O?
[Please use the second set on the right.]

19. What is the potential energy of the system at a point between A and O?
[Please use the second set on the right.]

20. Can the total energy of the system be all kinetic?
[Please use the *first* set.]

21. Can the total energy of the system be all potential?
[Please use the *first* set.]

22. Can the total energy of the system be kinetic and potential?
[Please use the *first* set.]

23. Can the total energy of the system (spring–mass) be negative if we ignore the gravitational potential energy of the mass?
[Please use the *first* set.]

24. Is the period of oscillation the same as the period of kinetic energy?
[Please use the *first* set.]

25. Does the frequency of oscillation of a spring depend upon the amplitude?
[Please use the *first* set.]

26. When we order springs from spring catalogs, do we generally do the calculations and order a spring of a particular spring constant?
[Please use the *first* set.]

27. Is the unit of spring constant newtons per meter (= N ∕ m)?
[Please use the *first* set.]

28. Is hertz (the unit of frequency) = 1 ∕ s?
[Please use the *first* set.]

29. Are there two similar points on either side of the equilibrium point that give the same value of kinetic energy in joules?
[Please use the *first* set.]

30. Are there two similar points on either side of the equilibrium point that give the same value of potential energy in joules?
[Please use the *first* set.]

31. If energy is conserved, will all points have the same total energy while the system is oscillating?
[Please use the *first* set.]

For problems **32-41**, a spring of force constant k on a smooth horizontal table is attached to a fixed support and is stretched horizontally through a distance $x_{DISPLACEMENT}$ along $+x$ from the spring's equilibrium position with a mass m at the end.

32. What is the force on the mass m due to the spring?

A. $k \times x$ opposite to the direction of stretching.
B. $k \times x$ in the direction of stretching.
C. Independent of the direction of stretch x.
D. $k \times x^2$ opposite to the direction of stretching.

33. What is its period of simple harmonic oscillation in seconds?

A. $2\pi \times$ square root of $[(2 \times k)/m]$.
B. $2\pi \times$ square root of (m/k).
C. $2\pi \times$ square root of (k/m).
D. $2\pi \times$ square root of $[m/(2 \times k)]$.

34. If the mass is released from rest after stretching the spring through a distance $A_{AMPLITUDE}$, what is the amplitude of the oscillation?

A. $2 \times A$.
B. $A/2$.
C. A.
D. $A/4$.

35. What is the expression for the velocity of the mass when the stretching is x from the equilibrium point?

A. $(\frac{1}{2} \times k \times A^2) - (\frac{1}{2} \times k \times x^2) = \frac{1}{2} \times m \times v^2$.
B. $A \times (k/m) \times [1 - (x^2/A^2)]$.
C. $A \times (k/m) \times x$.
D. $A \times (k/m)$.

36. What is the total energy of the oscillating spring–mass system when the stretching is $x_{DISPLACEMENT}$? Assume A is the amplitude and k is the spring constant.

   A. $\frac{1}{2} \times k^2 \times A^2$.
   B. $\frac{1}{2} \times k \times A^2$.
   C. $(k \times A)/4$.
   D. $k \times A^2$.

37. What is the kinetic energy when the displacement is x?

   A. $\frac{1}{2} \times k \times A^2$.
   B. $\frac{1}{2} \times k \times x^2$.
   C. $(\frac{1}{2} \times k \times A^2) - (\frac{1}{2} \times k \times x^2)$.
   D. $\frac{1}{2} \times m \times A^2 \times x^2$.

38. What is the potential energy when the spring is displaced through x from the equilibrium point?

   A. $\frac{1}{2} \times k \times A^2$.
   B. $\frac{1}{2} \times m \times A^2 \times x^2$.
   C. $(k \times x^2)/4$.
   D. $\frac{1}{2} \times k \times x^2$.

39. When the mass oscillates from the position of maximum displacement towards the equilibrium point, the potential energy of the spring changes from a maximum to zero but never negative.

   A. True.
   B. False.

40. Is the total energy of the system conserved if there is no friction?

   A. Yes.
   B. No.

41. Is the elastic force of the spring on the mass a constant within the elastic limits as it completes one full oscillation?

   A. Yes.
   B. No.

For problems **42-48**, a spring of spring constant k is vertically attached at Q and when a mass m is suspended at the other end B it stretches through a length l. When the mass is pulled down further through a distance A and released it executes simple harmonic oscillation.

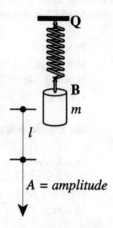

Problems 42-48

42. What is the spring constant of the spring?

    A. $(m \times g)/A$.
    B. $(m \times g)/l$.
    C. m/l.
    D. m/A.

43. What is the period of vertical oscillations of the mass?

    A. $2\pi \times$ square root of (A/g).
    B. $2\pi \times$ square root of $[(2 \times A)/g)]$.
    C. $2\pi \times$ square root of (l/g).
    D. $2\pi \times$ square root of $[(2 \times l)/g)]$.

44. What is the amplitude of vertical oscillation?

    A. l.
    B. $2 \times l$.
    C. $2 \times A$.
    D. A.

45. What is the frequency of oscillation of the mass?

    A. $(1/2\pi) \times$ square root of (g/l).
    B. $2\pi \times$ square root of (g/l).
    C. $(1/2\pi) \times$ square root of (l/g).
    D. $2\pi \times$ square root of (l/g).

46. What is the frequency of variation of kinetic energy?

    A. $(1/2\pi) \times$ square root of (g/l).
    B. $(1/\pi) \times$ square root of (g/l).
    C. $4\pi \times$ square root of (g/l).
    D. $\pi \times$ square root of (g/l).

47. What is the frequency of variation of elastic potential energy?

    A. $(1/2\pi) \times$ square root of (g/l).
    B. $2\pi \times$ square root of (g/l).
    C. $(1/\pi) \times$ square root of (g/l).
    D. $2\pi \times$ square root of (l/g).

48. What is the frequency of variation of total energy?

    A. $(1/2\pi) \times$ square root of (g/l).
    B. $(1/\pi) \times$ square root of (g/l).
    C. $2\pi \times$ square root of (l/g).
    D. Zero.

For problems **49-51**, two springs s_1 and s_2 of spring constants k_1 and k_2 are connected as shown in the figure to a mass m.

49. What is the force serving to stretch each spring?

    A. $m \times g$.
    B. $(m \times g)/2$.
    C. $(m \times g)/(k_1 + k_2)$.
    D. $(m \times g)/(k_1 - k_2)$.

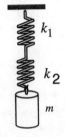

Problems 49-51

50. What is the total stretching of the springs?

    A. $(m \times g)/k_1$.
    B. $(m \times g)/k_2$.
    C. $(m \times g)/(k_1 + k_2)$.
    D. $[(m \times g)/k_1] + [(m \times g)/k_2]$.

51. What is the effective spring constant of the system?

    A. $k_1 + k_2$.
    B. $(k_1 \times k_2)/(k_1 + k_2)$.
    C. $k_1 - k_2$.
    D. $(k_1 + k_2)/2$.

For problems **52-55**, two springs s_1 and s_2 of spring constants k_1 and k_2 are connected as shown in the figure to a mass m.

52. If l is the effective stretching of the system, what is the stretching of each spring? Assume $k_1 = k_2$.

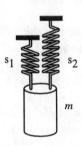

Problems 52-55

    A. $l/2$.
    B. l.
    C. $2(k_1/k_2) \times l$.
    D. $3(k_2/k_1) \times l$.

53. What is the force on springs s_1 and s_2 respectively? Assume the mass m is symmetrical, i.e. the center of mass is midway between the attachment points. Assume $l_2 = l_1 = l$ is the distance each spring is stretched when the mass is attached to the system.

    A. $(k_1 / k_2) \times l$ each.
    B. $k_2 \times l$ and $k_1 \times l$.
    C. $k_1 \times l$ and $k_2 \times l$.
    D. $(k_1 / k_2) \times l$ and $(k_2 / k_1) \times l$.

54. What is the effective spring constant if $l_2 = l_1 = l$?

    A. $k_1 + k_2$.
    B. $(k_1 \times k_2) / (k_1 + k_2)$.
    C. $(k_1 + k_2) / 2$.
    D. $2 \times (k_1 + k_2)$.

55. What is the stretching l of the system?

    A. $[(m \times g) / k_1] + [(m \times g) / k_2]$.
    B. $(m \times g) / k_1$.
    C. $(m \times g) / k_2$.
    D. $(m \times g) / (k_1 + k_2)$.

For problems **56 – 57**, a mass m is attached to a system of two vertical springs each of spring constant k.

56. What is the stretching of the system when the two identical springs are in series?

    A. $(2 \times m \times g) / k$.
    B. $(m \times g) / k$.
    C. $(m \times g) / (2 \times k)$.
    D. $(m \times g) / (4 \times k)$.

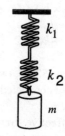

Problems 56-57

57. What is the stretching of the system when the two identical springs are in parallel?

    A. $(2 \times m \times g) / k$.
    B. $(m \times g) / k$.
    C. $(m \times g) / (2 \times k)$.
    D. $(m \times g) / (4 \times k)$.

58. A spring mattress has 16 springs of spring constant k. What is the frequency of vertical oscillation of a child of mass m on the mattress? Assume that all the springs are compressed equally at all times, as if the child were standing on a rigid board placed on top of the entire mattress.

    A. $2/\pi \times$ square root of (k/m).
    B. $16 \times$ square root of (k/m).
    C. $16/\pi \times$ square root of (k/m).
    D. $1/(8\,\pi) \times$ square root of (k/m).

For problems **59 – 60**, the system of two springs with spring constants k_1 and k_2 and a mass m is stretched between points P and Q.

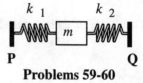

Problems 59-60

59. What is the force on m when it is displaced by distance x to the right of the equilibrium point?

    A. $(k_1 + k_2) \times x$ to the right.
    B. $(k_1 + k_2) \times x$ to the left.
    C. $(k_1 - k_2) \times x$ to the right.
    D. $(k_1 - k_2) \times x$ to the left.

60. The period of horizontal oscillation of the mass is:

    A. $2\,\pi \times$ square root of $\{[m \times (k_1 + k_2)]/(k_1 \times k_2)\}$.
    B. $2\,\pi \times$ square root of (m/k_1).
    C. $2\,\pi \times$ square root of $[m/(k_1 + k_2)]$.
    D. $2\,\pi \times$ square root of (m/k_2).

61. A spring–mass system is a simple harmonic motion. In the choices below, what else is simple harmonic motion?

    A.   A person sliding down an inclined plane.
    B.   A plate that falls to the ground.
    C.   Stretching a rubber band *beyond* the elastic limits.
    D.   The oscillation of a chandelier in the wind.

For problems **62 – 63**, a spring of spring constant k and length L stretches by a length l when a mass m is suspended from its end. The spring is folded and the mass m is suspended again.

62.   The effective spring constant of the folded spring is:

A. $2 \times k$.
B. $k / 2$.
C. k.
D. $4 \times k$.

63.   What is the stretching of the folded system?

A. $l / 4$.
B. $4 \times l$.
C. l.
D. $2 \times l$.

64.   Two masses m_1 and m_2 are connected by a spring of spring constant k and the system is placed on a horizontal frictionless table. What is the period of horizontal oscillation of each mass?

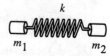

Problem 64

A. $2\pi \times$ square root of $[(m_1 + m_2) / k]$.
B. $2\pi \times$ square root of $\{[(m_1 \times m_2)] / [(m_1 + m_2) \times k]\}$.
C. $2\pi \times$ square root of (m_1 / k).
D. $2\pi \times$ square root of (m_2 / k).

65.   The mass of a chlorine atom is 35 times the mass m of a hydrogen atom. If the vibrational (stretching) frequency of an HCl molecule is n, what is the force constant k of the system in newtons/m?

A. $(35 / 9) \times \pi^2 \times n^2 \times m$.
B. $35 \times \pi^2 \times n^2 \times m$.
C. $(n^2 \times m) / 35$.
D. $35 \times n^2 \times m$.

66.   A mass m released from rest falls through a height h onto a spring on the ground. If the spring is compressed through a length s before the mass m comes to rest, what is the spring constant of the spring?

A. $[2 \times m \times g \times (h + s)] / s^2$.
B. $(m \times g \times h) / s^2$.
C. $(m \times g \times h) / (2 \times s^2)$.
D. $(m \times g \times h) / h^2$.

67. A light spring is kept stretched in the vertical position by suspending a mass m at the free end. When an additional mass $2 \times m$ is suspended, the spring is stretched by a length l. What is the period of oscillation of the system?

    A. $2\pi \times$ square root of (l/g).
    B. $2\pi \times$ square root of $[(2 \times l)/(3 \times g)]$.
    C. $2\pi \times$ square root of $[(2 \times l)/g]$.
    D. $2\pi \times$ square root of $[(3 \times l)/(2 \times g)]$.

For problems **68–69**, a particle of mass m_{MASS} is executing simple harmonic motion about a point O between points A and B. At a certain instant in time, the particle is at point P at a distance $x_{DISPLACEMENT}$ from the equilibrium point O. Assume the spring constant is $k_{\text{spring constant}}$. Ignore gravitational force. There is no frictional energy loss. Choose one answer:

    A.   Yes! Acceleration and velocity vectors are in the same direction. This will increase the speed.
    B.   No. Acceleration and velocity vectors are in the opposite directions. This will reduce the speed.

68. Are the velocity and acceleration vectors in the same direction at point P if the mass is going towards the end point (of maximum displacement) B from the equilibrium point O?

69. Are the velocity and acceleration vectors in the same direction at point P if the mass is going towards the equilibrium point O from the end point (of maximum displacement) B?

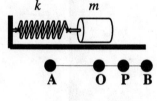

Problems 68-69

ZONE *solution*

1.   Let t_1 be the time to go from A to B, and t_2 the time to come back from B to A.
     The total time for the trip is the sum: $t_1 + t_2$.
     Average speed = distance / time.
     $\Rightarrow$ *Time = distance / average speed* if speed is a constant.
     This gives us: $t_{1,\text{ TIME IN s}} = d_{\text{ DISTANCE IN m}} / v_{1,\text{ SPEED IN m/s}}$,
     $t_{2,\text{ TIME IN s}} = d_{\text{ DISTANCE}} / v_{2,\text{ SPEED IN m/s}}$.
     $t_{\text{ TOTAL TIME IN s}} = t_1 + t_2$
     $= (d / v_1)_{\text{TIME TO GO FROM A TO B}} + (d / v_2)_{\text{TIME TO COME BACK FROM B TO A}}$.
     Answer B!

2.   You are correct! Distance is a scalar, so we add:
     $d_{\text{ TOTAL DISTANCE IN m}} = d_{\text{ A to B IN m}} + d_{\text{ B TO A IN m}} = 2d$ meters.
     Answer A!

4.   *Average speed = <u>total</u> distance / time.*
     (1)   Distance is a scalar, so we add: $d_{\text{ TOTAL}} = d + d = 2d$.
     (2) $t_{1,\text{ TIME IN s}} = d_{\text{ DISTANCE IN m}} / v_{1,\text{ SPEED IN m/s}}$, and
           $t_{2,\text{ TIME IN s}} = d_{\text{ DISTANCE}} / v_{2,\text{ SPEED IN m/s}}$.
     The total time is:
     $t_{\text{ TOTAL TIME IN s}} = t_1 + t_2 =$
     $(d / v_1)_{\text{ TIME TO GO FROM A TO B}} + (d / v_2)_{\text{ TIME TO COME BACK FROM B TO A}}$.
     Average speed = total distance / time.
     $\Rightarrow v = (2d)_{\text{ TOTAL DISTANCE IN m}} / (t_1 + t_2)$
     $= (2d)_{\text{ TOTAL DISTANCE IN m}} / [(d / v_1) + (d / v_2)]_{\text{ TOTAL TIME IN s}}$
     $= 2 / [(1 / v_1) + (1 / v_2)] = 2 / [(v_1 + v_2) / (v_1 v_2)]$
     Play your math. $v_{\text{ AVERAGE SPEED IN m/s}} = (2 \times v_1 \times v_2) / (v_1 + v_2)$.

     Answer B!

5.   *Average velocity = displacement / time.*
     Oops! Displacement is a vector:
     $d_{\text{ TOTAL DISPLACEMENT IN m}} =$
           $[d(\text{along } +x)]_{\text{DISPLACEMENT A TO B IN m}} + [d(\text{along } -x)]_{\text{DISPLACEMENT B TO A IN m}}$.
           $\Rightarrow d + (-d) = 0$.
     That is, the net displacement is zero when you are back at the point you started, point A.
     $\Rightarrow$ *Average velocity = displacement between starting and finishing point / time.*
     $\Rightarrow v_{\text{ AVERAGE VELOCITY IN m/s}} = (0 - 0)_{\text{ METERS}} / t_{\text{ SECONDS}} = 0$ m/s!
     I am sure you will never forget that displacement is a vector.
     Answer D!

7. If speed is constant then *time* = *distance / average speed.*
Distance for the *first* segment = $d / 4$.
Average speed for the *first* segment = v_1.
Distance for the *second* segment = $d / 4$.
Average speed for the *second* segment = v_2.
Distance for the *third* segment = $d / 2$.
Average speed for the *third* segment = v_3.
The time for the first segment of the trip is:

$t_{1, \text{TIME FOR FIRST SEGMENT IN s}} =$
$[(d / 4)_{\text{DISTANCE IN m}} / v_{1, \text{AVERAGE SPEED IN m/s}}]_{\text{FIRST SEGMENT}} = d / (4 \times v_1)$.
Similarly $t_{2, \text{TIME FOR SECOND SEGMENT IN s}}$
$= [(d / 4)_{\text{DISTANCE}} / v_{2, \text{AVERAGE SPEED}}]_{\text{SECOND SEGMENT}} = d / (4 \times v_2)$.
$t_{3, \text{TIME FOR THIRD SEGMENT}} = [(d / 2) / v_3]_{\text{THIRD SEGMENT}} = d / (2 \times v_3)$.
The average speed for the entire trip is given by *total distance / total time*, or
$v_{\text{AVG}} = d_{\text{TOTAL}} / (t_1 + t_2 + t_3)$.
$\Rightarrow v_{\text{AVERAGE SPEED FOR THE ENTIRE TRIP}} =$
$[(d / 4) + (d / 4) + (d / 2)] / \{[d / (4 \times v_1)] + [d / (4 \times v_2)] + [d / (2 \times v_3)]\}$.
$= d / \{[d / (4 \times v_1)] + [d / (4 \times v_2)] + [d / (2 \times v_3)]\}$. Cancel d through out.
$\Rightarrow v = 1 / \{[1 / (4 \times v_1)] + [1 / (4 \times v_2)] + [1 / (2 \times v_3)]\}$.
$\Rightarrow v = (4 \times v_1 \times v_2 \times v_3) / [(v_2 \times v_3 + v_1 \times v_3) + (2 \times v_1 \times v_2)]$.
   [Note: $[1 / (4 \times v_1)] + [1 / (4 \times v_2)] = (v_2 + v_1) / (4 \times v_1 \times v_2)$.
   $[(v_2 + v_1) / (4 \times v_1 \times v_2)] + [1 / (2 \times v_3)]$
   $= [1 / (4 \times v_1 \times v_2 \times v_3)] \times \{[(v_2 + v_1) \times v_3] + (2 v_1 v_2)\}$
   $= [1 / (4 \times v_1 \times v_2 \times v_3)] \times [(v_2 v_3 + v_1 v_3) + (2 v_1 v_2)]$. I am sure you know the rest.]
Answer A!

8. We hope you did not get too tired after doing problem **7**. Let us do this one.
$$Distance = Average\ speed \times time.$$
The distances traveled in the first half and second half of the trip are:

$d_{1, \text{DISTANCE FOR THE FIRST HALF OF TIME}} = [v_{1, \text{AVERAGE SPEED}} \times (t_{\text{TIME FOR FULL TRIP}} / 2)]$.
$d_{2, \text{DISTANCE FOR THE SECOND HALF OF TIME}} = [v_{2, \text{AVERAGE SPEED}} \times (t_{\text{TIME FOR FULL TRIP}} / 2)$ respectively.
The average speed is $v_{\text{AVG}} = d_{\text{TOTAL}} / t_{\text{TOTAL}}$.
   $\Rightarrow v_{\text{AVG}} = (d_1 + d_2)_{\text{TOTAL DISTANCE}} / t_{\text{TOTAL TIME}}$.
$v_{\text{AVG}} = \{[v_1 \times (t / 2)] + [v_2 \times (t / 2)]\} / t$. Cancel time t through out.
$v_{\text{AVG}} = (v_1 / 2) + (v_2 / 2) = (v_1 + v_2) / 2$.
Answer A!

10. Thank God that this is a qualitative problem.
When you throw a ball vertically up, its velocity is instantaneously zero at the *maximum* height even though acceleration is non-zero.
[Think about this: If the ball's velocity and acceleration were both zero at the maximum height, its speed

would have remained at the constant value of zero forever. The body would never come down. Things at rest stay at rest! We know that this is not the case. That is, acceleration at maximum height is *not* zero.]
Answer A!

11.   When you go at a non-zero speed on a highway, and do not change the speed or the direction, acceleration is zero even though velocity is not. That is, if velocity is a constant then acceleration which is the *rate of change* of velocity will also be zero. Another example of zero acceleration and non-zero velocity is your garage door going up vertically!
Answer A!

13.   When you "step" on the gas, velocity and acceleration are in the same direction. Speed increases. This is acceleration.
When you drop something, both velocity and acceleration are in the same direction! Speed increases. This is also an example of acceleration.
[Note: When you throw something vertically up, velocity and acceleration are not in the same direction! They are actually in the opposite directions. This is defined as deceleration. Speed decreases.]
Answer A!

14.   Total distance traveled is half the circumference $= 2\pi R / 2 = \pi R$.
This is the arc length from point A to point B.
Net displacement is twice the radius or $2R$. Note that the displacement is a vector.
Answer D!

16.   Since the velocity is not a constant (speed is but direction is not!), the acceleration will be non-zero. There is a change in the direction.
Velocity is a vector. It has magnitude (which is the instantaneous speed) and direction. If either magnitude or direction changes, then its velocity is <u>not</u> a constant.
Answer C!

17.   75 miles/hour $\times$ (1 hour/ 3600 seconds) $\times$ (1,610 m/mile) $\approx$ 33 m/s.
When you go on a highway with this kind of speed, you are covering 33 m in one second. Do not forget that. Be very attentive! (How many times have we heard that?)
Answer A!

19.   Answer A!

20.   *Speed = distance / time.*
$\Rightarrow$ Time   = (distance to the sun / speed of light)
                  = 150×10^9 m / $(3 \times 10^8$ m/s)
                  = 500 seconds = 8.3 minutes.

(Note: The light that you see at 9.30 AM is actually produced 8.3 minutes before at the sun if you ignore the scattering of light. At your level, you can ignore learning about scattering.)
Answer B!

22. False! Speed of an object along $-y$ increases by 10 m/s, every second because acceleration due to gravity on the surface of the earth is 10 m/s per second along $-y$!
     (Do not forget this: As speed increases, you cover *more* distance per unit time.)
     Answer B!

23. Answer A!

25. Answer A!

26. *Average speed* $\approx$ *distance / time.*
     Average speed $\approx$ *(circumference of the orbit) / (time for the earth to complete one rotation about the sun).*

     Circumference of the orbit
     $= 2 \times \pi \times R$ where R is the distance to the center of the sun from the center of the earth $= 150 \times 10^9$ m.
     Time for the earth to complete one rotation about the sun = 1 year
             = 1 year $\times$ (365 days / 1 year) $\times$ (24 hours / 1 day) $\times$ (3,600 s / 1 hour).

     $\Rightarrow v = (2 \times \pi \times R / 1 \text{ year}) = (2 \times \pi \times 150 \times 10^9 \text{ m}) / (365 \times 24 \times 3600 \text{ s}) = 29.9 \times 10^3$ m/s
     = 30,000 m/s.
     That is real "fast" compared to the speed of an airplane. An airplane can travel at 220 m/s.   Answer B!

28. <u>To find average velocity:</u>
     Average velocity = displacement / (time to go from A to C)
                  $= (\text{distance AC}) / 80 \text{ min} = (40^2 + 40^2)^{1/2}$ m / 80 min
                  = (0.7 m / min) $\times$ (60 min / 1 hr) = 42 m/hr <u>along AC.</u>
     If you are confused about the direction, here is more help:
     To be precise, the direction of a vector is determined by:
     tan ϕ = (*y component*) /(*x component*).
     tan ϕ = 40 / 40 = 1 or ϕ = 45 degrees with respect to *x* axis or 45 degrees above the east.

     <u>To find the average speed:</u>
     The total distance traveled = 40 m + 40 m = 80 m.
     Total time = 80 min.
     Average speed = (total distance / time) = 80 m / 80 min = 1 m / min
     = (1 m / minute) $\times$ (60 min / 1 hr) = 60 m/hr.
     Answer D!

29. A bus can maintain a constant speed of 65 mi/hr. It may or may not be going only in a certain direction. Even freeways have curves! When direction is not a constant, velocity is not a constant. When velocity is changes, we have acceleration.
Answer B!

31. Since you are told that the velocity is a constant, the change in velocity is zero, so the acceleration is zero. Note that *average acceleration = change in velocity / time.* If *change* in velocity is zero, acceleration is always zero.
Answer C!

32. For objects moving in the downward direction, acceleration due to gravity increases the speed (Here acceleration and velocity have the same direction.), so the velocity is increased in magnitude. Note that the acceleration due to gravity remains a constant near the surface of earth!
Answer E!

34. **Common Sense:** During free fall downward, *speed* increases by 10 m/s for every second.
30 m/s + 10 m/s = 40 m/s.
Formula Pundit: Use $v_{fy} = v_{0y} + a_y t$

$[v_{fy}$	=	final vertical component of the velocity vector in m/s,
v_{0y}	=	initial vertical component of the velocity vector = $-$ 30 m/s,
a_y	=	the acceleration due to gravity in m/s/s in the vertically down direction
	=	$-$ 10 m/s/s,
t	=	the time interval in s = 1 s.]

or $v_{fy} = (-30) + [(-10) \times 1] = -40$ m/s.
The negative sign indicates a motion in the $-y$ direction (down).
Answer C!

35. Because speed increases by 10 m/s for a rose that is dropped, distance traveled in the *second* second will be more than the distance traveled in the first second. The distance traveled in the third second will be even greater.

(1) Speed at the end of first second = 10 m/s. The average speed in the first second is
(0 + 10) / 2 = 5 m/s. *Distance = average speed × time* = 5 m/s × 1 s = 5 m.

(2) Speed at the end of *second* second = 20 m/s. The average speed for the first two seconds =
(0 + 20) / 2 = 10 m/s. *Distance = average speed × time* = 10 m/s × 2 s
= 20 m.

(3) Speed at the end of *third* second = 30 m/s. The average speed between start and finish = (0 + 30) /
2 = 15 m/s. *Distance = average speed × time* = 15 m/s × 3 s
= 45 m.

Formula Pundit: $\Delta y = (v_{oy} \times t) + [(1/2)\, a_y\, t^2]$

[$\Delta y (= y_f - y_0)$ is the vertical displacement in m $= (y_f - 0\;_{\text{STARTING POSITION}})$,
v_{0y} is the initial vertical component of the velocity vector $= 0$ m/s,
a_y is the acceleration due to gravity in m/s/s in the vertically down direction $= -10$ m/s/s,
t is the time interval in s.]

$(y_f - 0) = (0 \times t) + [(1/2) \times (-10) \times t^2]$ or $y_f = (1/2) \times (-10) \times t^2$.
When $t = 1$, $y_f = -5$ m.
When $t = 2$ s, $y_f = -20$ m.
When $t = 3$ s, $y_f = -45$ m.
The negative sign indicates that the position of the rose is below the origin, $y = 0$ m.
Answer C!

37. Acceleration is defined as change in velocity per unit time. If unit time is chosen as 1 s, acceleration is change in velocity per second.
Answer D!

38. Definitely the one with the maximum acceleration. It gets your car to a higher speed (covering more distance per unit time) more quickly. Sports cars do just that!
Answer A!

40. **Common Sense:** Speed is continuously changing at the rate of 5 m/s every second.
Speed at the end of 5 s $= 0\;_{\text{SPEED AT THE START}} + 5 + 5 + 5 + 5 + 5 = 25$ m/s!
Formula Pundit: $v_f = v_0 + at$.

[$v_f =$ the final velocity in m/s, $v_0 =$ the initial velocity in m/s $= 0$ m/s,
$a =$ the acceleration in m/s/s $= 5$ m/s/s, $t =$ the time interval in s $= 5$ s.]

$v_f = 0 + (5 \times 5) = 25$ m/s if the car is moving to the right (-25 m/s if the car is moving to the left).
Answer C!

41. **Common Sense:** Speed is continuously changing at the rate of 3 m/s every second.
Speed at the end of 5 s $= 4\;_{\text{SPEED AT THE START}} + 3 + 3 + 3 + 3 + 3 = 4 + 15 = 19$ m/s!

Formula Pundit: $v_f = v_0 + at$.
[$v_f =$ the final velocity in m/s, $v_0 =$ the initial velocity in m/s $= 4$ m/s,
$a =$ the acceleration in m/s/s $= 3$ m/s/s, $t =$ the time interval in s $= 5$ s.]

$v_f = 4 + (5 \times 3) = 19$ m/s if the car is moving to the right (-19 m/s if the car is moving to the left).
Answer C!

43.   Imagine moving in circles. The direction changes. Velocity has magnitude and direction. If direction of motion changes, there is a change in velocity!
Answer A!

44.   Since the vectors in the problem have the same magnitude but *different* directions, they are equally big.
Answer C!

46.   The slope of the line is a constant.
Slope = Δs WHAT IS PLOTTED ON **y** / Δt WHAT IS PLOTTED ON **x**
= displacement / time = velocity. If slope is a constant, velocity also is a constant.
Answer C!

47.   Displacement–time graph is a curve, and it looks like the graph of $(1/2) \times at^2$. Here "a" is for acceleration. [Is it accelerating or decelerating? You and I can agree that the slope (velocity) is positive. Also we see that the slope is getting bigger. This makes acceleration which is the rate of change of velocity positive in sign. Since both velocity and acceleration have the same sign, speed has to increase. That is it is accelerating.]
Answer B!

49.   Since acceleration is the *rate of change* of velocity, we need to know the velocity–time graph to find acceleration.
Answer B!

50.   Use $v_{fy} = v_{0y} + a_y t$.

[v_{fy} = the final vertical component of the velocity vector in m/s,
v_{0y} = the initial vertical component of the velocity vector = 0 m/s,
a_y = the acceleration due to gravity in m/s/s in the vertically down direction
= -10 m/s/s,
t = the time interval in s.]

If the book is dropped from rest, initial velocity v_{0y} will be zero.
This gives us $v_{fy} = v_{0y} + a_y t = 0 + (-10)t = -10t$.
If the final velocity v_{fy} is plotted on **y** and time t on **x**, then the graph will be a straight line with a negative slope.
Answer D!

52.   As the person increases the speed to a higher value, CB represents the final speed attained, at point B.
Answer D!

CHAPTER 1, S o l u t i o n P a g e **7**

53.   Whenever a graph is plotted, *tan θ* always represents the slope $(v_f - v_0) / t$.
From a practical conceptual point of view, it represents:

$tan \theta = (\Delta y$ WHAT IS PLOTTED ON **y** AXIS$) / (\Delta x$ WHAT IS PLOTTED ON **x** AXIS$)$.
This is nothing but acceleration, as $a = \Delta v / \Delta t$.
Answer E!

55.   When a ball is thrown vertically up, first of all it continually reduces the speed due to displacement (or for that matter velocity) and acceleration being in the opposite directions.
At the maximum height, it comes to a zero instantaneous velocity.
From then on, as the ball comes back towards the ground, acceleration and velocity are in the same direction (they have the same sign), and so it speeds up on the way down. It accelerates. If "up velocity" is positive, then "down velocity" is negative.
Answer B!

56.   Speed is *always* positive. That means we can neglect figures which go below the **x** axis.
On the way up, speed *reduces*, becomes a zero at the maximum height, and then increase on the way down due to a non-zero acceleration due to gravity. Whatever figures have increasing speed from the start, we can neglect it.
In Figure **C**, speed increases from the start which is not the case. So we neglect Figure **C**.
Answer D!

58.   The area of velocity-time graph is nothing but displacement.
Answer B!

59.   Anything that is thrown off a moving object has the velocity of the moving object at the start. Since the helicopter is initially going up, the cigarette butt will also have the same initial velocity along the "up."
Answer A!

61. $v_{fy} = v_{0y} + a_y \times t.$ **(1)**
$v_{fy}^2 = v_0^2 + 2 \times a_y \times (\Delta y).$ **(2)**
$\Delta y = v_{0y} \times t + (1/2) \times a_y \times t^2.$ **(3)**

$[\Delta y (= y_f - y_0)$ is the vertical displacement in m $= (h - 0) = h$ m,
v_{0y} is the non-zero initial vertical component of the velocity vector,
v_{fy} is the final vertical component of the velocity vector $= 0$ m/s at the maximum height,
a_y is the acceleration due to gravity in m/s/s in the vertically down direction $= -10$ m/s/s,
t is the time interval in s.]

At maximum height, final speed along the vertical *y* is zero. Substituting into equation **(2)**, we get;
$0 = v_{0y}^2 + 2 \times (-g) \times h. \Rightarrow v_{0y}^2 = 2 \times g \times h. \Rightarrow h = v_{0y}^2 / (2 \times g).$
Answer B!

62.   If initial and final positions are the same, displacement will be: $y_f - y_i = 0$ m as initial and final points are the same. Note that displacement along the horizontal direction x is also zero as motion is strictly along the vertical.
Answer C!

64.   Time to reach the ground is $2 \times$ the time to reach the maximum height.
Use equation (1) of problem **61**. $v_{fy} = v_{0y} + (a_y \times t)$.
At maximum height, $v_{fy} = 0$ m/s.
$\Rightarrow 0 = v_{0y} + (-g) \times t$.
Solve for t.
Time to reach the maximum height: $t = v_{0y} / g$.
$\Rightarrow$ Total time is thus twice this $= 2 \times (v_{0y} / g)$.
Answer A!

65.   If gravity (the acceleration) is turned off at the maximum height, object maintains zero speed along y. It just stays put! We expect the velocity to remain the same at 0 m/s.
Answer B!

67.   Velocity and acceleration can have opposite signs.
Answer D!

68.   *Average speed = distance / time.*
The total length of the train has to get past the bridge.
This makes the total distance $= L$ LENGTH OF THE BRIDGE $+ l$ LENGTH OF THE TRAIN.
Since speed is a constant, we can use:
t TIME IN s $= (L + l)$ DISTANCE IN m $/ v$ AVERAGE SPEED IN m/s.
Answer B!

70.   The trick is to understand regions of deceleration where you have to put a negative sign for "a" in the equations of kinematics.
For regions of constant acceleration: Δx is proportional to $(1/2) \times a_y \times t^2$.
Deceleration a_y carries a negative sign. When we look at the graph we see the speed increasing (region of constant acceleration as slope is a constant), attaining a constant value, speed decreasing, becoming a constant, and then finally increasing.
Misconception: Just because the "curve" is going to the right does not mean that the particle is always moving to the right. Whenever the particle moves to the left, velocity is negative.

(1)   From $t = 0$ to $t = 2$ s, velocity is increasing. Since the slope is a constant, we conclude that it is accelerating. So:
$\Delta x = (1/2) \times a$ ACCELERATION $\times t^2 = (1/2) \times [(v_f - v_0) / \Delta t]$ ACCELERATION $\times t^2$.
$= (1/2) \times [(2 \text{ m/s} - 0 \text{ m/s}) / 2 \text{ s}] \times (2 \text{ s})^2 = 2$ m.

CHAPTER 1, Solution Page **9**

(2)   From $t = 2$ s to $t = 6$ s, the horizontal line tells us that speed is a constant. Use:
$\Delta x = v \times t$ at that point: $\Delta x = (2 \text{ m/s} \times 4 \text{ s}) = 8$ m.

(3)   From $t = 6$ s to $t = 10$ s, velocity is on the decrease (decelerating). Use:
$\Delta x = (v_{0x} \times t) + (1/2) \times a \times t^2$
$= (2 \text{ m/s} \times 4 \text{ s}) + \frac{1}{2} \times [(-2 \text{ m/s} - 2 \text{ m/s}) / 4 \text{ s}] \times (4 \text{ s})^2 = 8 \text{ m} + (-8 \text{ m}) = 0$ m.

(4)   From $t = 10$ s to $t = 14$ s, the object travels at constant velocity but in the opposite direction (velocity is negative). $\Delta x = (-2 \text{ m/s} \times 4 \text{ s}) = -8$ m.

Adding the displacements for various intervals, we get:

total displacement $= 2 + 8 + 0 + (-8) = 2$ m.

The object is now 2 m *in front* of the starting point.

Answer D!

71.   Acceleration, $a = \Delta v / \Delta t$ is zero when slope is zero, i.e. from A to B (or for that matter from C to D).
Answer B!

73. $\Delta y = (v_{0y} \times t) + [(1/2) \times a_y \times t^2]$.

[$\Delta y\ (= y_f - y_0)$ is the vertical displacement in m $= (-h - 0) = -h$ m,
v_{0y} is the non-zero initial vertical component of the velocity vector,
v_{fy} is the final vertical component of the velocity vector,
a_y is the acceleration due to gravity in m/s/s in the vertically down direction $= -10$ m/s/s,
t is the time interval in s.]

$\Rightarrow -h = (v_{0y} \times t) + [(1/2) \times (-g) \times t^2]$.
$\Rightarrow [h - (1/2) \times g \times t^2] / t = v_{0y}$.
Answer E!

74. $\Delta y = (v_{0y} \times t) + (1/2) \times a_y \times t^2$.

[$\Delta y\ (= y_f - y_0)$ is the vertical displacement in m $= (-h - 0) = -h$ m,
v_{0y} is the initial vertical component of the velocity vector $= 0$ m/s,
a_y is the acceleration due to gravity in m/s/s in the vertically down direction $= -10$ m/s/s,
t is the time interval in s.]

$-h = (0 \times t) + [(1/2) \times (-g) \times t^2]$.
$\Rightarrow -h = (1/2) \times (-g) \times t^2. \Rightarrow h = [(1/2) \times g \times t^2]$.
$\Rightarrow t = [(2 \times h) / g]^{1/2}$.
Answer A!

76. Note that angle ABF = 90°. Also, angle ACF = 90°.

The acceleration due to gravity acts along AF. If we want to know its component along AB, all we have to do is use the TRIG. This will give us acceleration along AC.

$cos\ \theta_1$ = Adjacent side to θ_1 / hypotenuse.

$\Rightarrow cos\ \theta_1$ = The component of "g" along AC / g .

The component of "g" along AC = acceleration along AC = $a_1 = g \times cos\ \theta_1$.

$a_1 = g \times cos\ \theta_1$ and $a_2 = g \times cos\ \theta_2$.

Answer B!

77. Note that angle ABF = 90°. Also, angle ACF = 90°.

The acceleration due to gravity acts along AF. If we want to know its component along AB, all we have to do is use the TRIG.

$cos\ \theta_1$ = Adjacent side to θ_1 / hypotenuse.

$\Rightarrow cos\ \theta_1$ = The component of "g" along AC / g .

The component of "g" along AC = acceleration along AC = $a_1 = g \times cos\ \theta_1$.

Distance, AC = $(1/2) \times a_1 \times t_1^2$ if released from rest.

$\Rightarrow t_1 = [2 \times (AC) / a_1]^{1/2}$.

Let us solve for $[(AC) / a_1]$.

In terms of angle, θ_1, and radius, R, AC = $2 \times R \times cos\ \theta_1$.

$\Rightarrow (AC) / a_1 = (2 \times R \times cos\ \theta_1) / (g \times cos\ \theta_1] = [(2 \times R) / g]$.

$\Rightarrow t_1 = [2 \times (AC) / a_1]^{1/2}$ becomes $t_1 = [2 \times (2 \times R) / g]^{1/2}$. This is independent of the angle! Interesting result?

Answer B!

79. Answer B is wrong because the correct unit of acceleration is in m/s/s.

Horizontal acceleration = 0 m/s/s.

Answer D!

80. For blue and black pens: Δy is the same. Initial vertical component of the velocity vector is 0 m/s. Their time is the same!

$\Delta y = (v_{0y} \times t) + (1/2) \times a_y \times t^2$.

$[\Delta y\ (= y_f - y_0)$ is the vertical displacement in m = $(-h - 0) = -h$ m,

v_{0y} is the initial vertical component of the velocity vector = 0 m/s,

a_y is the acceleration due to gravity in m/s/s in the vertically down direction = -10 m/s/s,

t is the time interval in s.]

The above expression reduces to: $\Delta y = (1/2) \times a_y \times t^2$.

Δy and a_y are the same for blue and black pens. That is, they strike the ground at the same time.

Answer C!

82.   Speed increases during downward fall because velocity and acceleration have the same (negative) sign. $\Rightarrow$
        Distance covered per unit time also increases.
        Answer A!

83.   **Common Sense:** Speed after 4 s is: $0 \text{ m/s} + (3 + 3 + 3 + 3) = 12 \text{ m/s}$.
        Average velocity $= [(v_f + v_0) \, / \, 2] = (12 + 0) \, / \, 2 = 6 \text{ m/s}$.
        Displacement $=$ (Average velocity) $\times$ time $= (6 \text{ m/s}) \times (4 \text{ s}) = 24 \text{ m}$.

        Formula Pundit: Use $\Delta x = (v_0 \times t) + (½ \times a \times t^2)$

        [$v_0 =$ initial velocity in m/s $= 0$ m/s, $a =$ acceleration in m/s/s $= 3$ m/s/s, $t =$ time interval $= 4$ s.]

        $\Delta x = 0 + [½ \times (3 \text{ m/s}^2) \times 4^2] = 24 \text{ m}$.
        Answer D!

85.   **Common sense:**   Speed after 1 s $= 10 - 3 = 7 \text{ m/s}$.
                                    Speed after 2 s $= 7 - 3$ in the *second* second $= 4 \text{ m/s}$.
        Formula: $v_f = v_0 + (a \times t) = 10 \text{ m/s} + (-3 \text{ m/s}^2) \times (2 \text{ s})$. $\Rightarrow v_f = 4 \text{ m/s}$.
        Answer A!

86. $v_f = v_0 + (a \times t)$.

        [$v_f =$ final velocity in m/s $= 4$ m/s, $v_0 =$ initial velocity in m/s, $a =$ acceleration in m/s/s
        $= -5$ m/s/s, $t =$ time interval in s $= 2$ s. Note the negative sign for acceleration. The car is decelerating.
        That means acceleration a and velocity have opposite signs.]

        $\Rightarrow 4 \text{ m/s} = v_0 + [-5 \text{ m/s/s} \times (2 \text{ s})]$. $\Rightarrow 4 \text{ m/s} = v_0 - 10$.
        $\Rightarrow v_0 = 10 + 4 = 14 \text{ m/s}$.
        Answer A!

88.   From 20 m/s, the chalk slows down to 10 m/s in one second as deceleration is 10 m/s per second. In
        another second, it has a speed of $10 + (-10)$ or zero.

        Use the formula: $v_{fy} = v_{0y} + (a_y \times t)$.

        [$v_{fy} =$ final vertical velocity at the maximum height in m/s $= 0$ m/s,
        $v_{0y} =$ initial vertical velocity in m/s $= 20$ m/s, $a_y =$ acceleration due to gravity in m/s/s
        $= -10 \text{ m/s}^2$, $t =$ time interval in s].

        $\Rightarrow 0 = 20 \text{ m/s} + [(-10 \text{ m/s}^2) \times t]$.
        $\Rightarrow t = 20 \, / \, 10 = 2 \text{ s}$.

Total time = 2 GOING UP + 2 COMING DOWN = 4 s.
Answer D!

89. **Common sense:** 10 m/s.
    Formula: $v_{fy} = v_{0y} + (a_y \times t)$.

    [v_{fy} = final vertical velocity at the maximum height in m/s = 0 m/s,
    v_{0y} = initial vertical velocity in m/s, a_y = acceleration due to gravity in m/s/s = -10 m/s^2,
    t = time interval in s = 1 s].

    $\Rightarrow 0 = v_{0y} + [(-10) \times (1 \text{ s})]$. $\Rightarrow v_{0y} = 10$ m/s.
    Answer A!

91. Use: $v_{fy}^2 = v_{0y}^2 + (2 \times a_y \times \Delta y)$.

    [v_{fy} = final vertical velocity, v_{0y} = initial vertical velocity in m/s = 0 m/s, a_y = acceleration due to gravity in
    m/s/s = -10 m/s^2, Δy = -1.8 m (landing point below the starting point)].

    $\Rightarrow v_{fy}^2 = 0 + [2 \times (-10) \times (-1.8)]$.
    $\Rightarrow v_{fy}^2 = 36$.
    Thus $v_{fy} = 6$ m/s or -6 m/s. Since the object is moving down we take the negative root if asked for the
    velocity just before impact.
    Answer C!

92. On the way down from maximum height:
    [v_{fy} = final vertical velocity, v_{0y} = initial vertical velocity in m/s = 0 m/s, a_y = acceleration due to gravity in
    m/s/s = -10 m/s^2, t = time interval in s = 1 s to go up (or 1 s to come down)].

    Time = 1 s up plus 1 s down.
    For 1 s down, $\Delta y = (v_{0y} \times t) + (\tfrac{1}{2} \times g \times t^2) = 0 + [\tfrac{1}{2} \times (-10) \times (1)^2] = -5$ m.
    Answer C!

94. Constant velocity means *change* of velocity is zero. $\Rightarrow$ Acceleration = 0. False.
    Answer B!

95. When you apply brakes, displacement (or for that matter velocity) and acceleration are in opposite directions.
    True.
    Answer A!

97. Time to go up is the same as time to come down if there is no air resistance. False.
    Answer B!

98. If I throw a ball up at 20 m/s, then on the way down, if I catch the ball at the same height, the velocity is -20 m/s if there is no air resistance. It is important stuff.

<u>Throwing up at 20 m/s to the maximum height:</u>

$v_{fy} = v_{0y} + (a_y \times t)$.

$\Rightarrow 0_{\text{VELOCITY AT MAX. HEIGHT}} = 20_{\text{INITIAL VELOCITY}} + [(-10) \times t]$. $\Rightarrow t = 2$ s.

<u>On the way down from maximum height:</u>

$v_{fy} = v_{0y} + (a_y \times t)$.

$v_{fy} = 0 + (-10 \times 2)$. $\Rightarrow v_{fy} = -20$ m/s.

If you throw something vertically up at 20 m/s, and if you catch it at the same height, then their speeds will have to be the same.

(The magnitudes of their velocities are the same but directions are not!)

Answer B!

100. Velocity is a vector. Let the wall be on the left. You throw the ball left towards the wall with a speed v but its velocity is $-v$ if $+x$ is towards the right. After it bounces, it has a velocity $+v$.

The change in velocity = $v_{\text{FINAL VELOCITY}} - (-v_{\text{INITIAL VELOCITY}}) = 2 \times v$.

Note that the change in speed = 0.

Answer B!

101. Yes for segments of constant acceleration.

Answer B!

103. $v_{avg} = (v_f - v_0) / t = (8 \text{ m} - 0) / (5 \text{ s} - 2 \text{ s}) = 8 / 3 = 2.7$ m/s.

Answer A!

104. a is constant but negative because slope is negative.

Answer B!

106. a is positive as slope is positive.

Answer A!

107. <u>On the way up:</u>

Speed reduces. Velocity is positive. Velocity is zero at the maximum height.

<u>On the way down from maximum height:</u>

Velocity is zero at maximum height. Speed continuously increases. Velocity is negative.

Answer B!

109. <u>On the way down:</u>
Velocity is non-zero and negative at the start. Speed continuously increases. Velocity is always negative.
Answer E!

110. Deceleration is zero if there is no friction. $\Rightarrow$ Velocity is a constant in time.
Answer A!

112. Velocity is a constant in time.
Answer A!

113. Use: $\Delta y = (v_{0y} \times t) + (\frac{1}{2} \times a_y \times t^2)$.

[v_{0y} = initial vertical velocity in m/s = 400 m/s, a_y = acceleration due to gravity in m/s/s = -10 m/s^2, t = time interval in s = 1 s].

$\Rightarrow \Delta y = [(400 \text{ m/s}) \times 1 \text{ s}] + [\frac{1}{2} \times (-10 \text{ m/s}^2) \times (1 \text{ s})^2] = [400 + (-5)]$ m.
Answer C!

1. $v_{0x} = v_0 \times \cos \alpha$.

HELP: v_{0x} = Initial *horizontal* velocity in m/s (the same as final horizontal velocity if horizontal acceleration is 0 m/s/s). v_0 = Projection speed (magnitude of the velocity vector at the start) in m/s. α = Projection angle with respect to the horizontal.

Real Beef: We often want horizontal and vertical components of a vector.

The vector by itself may not be useful.

Once we split the vector into its components, it is much easier to handle motion along two <u>independent</u> directions (horizontal *x* -, and vertical *y* -).

Brush up on your trig (that thing that you want to avoid learning):

$\cos \alpha$ = adjacent side to α / hypotenuse = v_{0x} / v_0. $\Rightarrow v_{0x} = v_0 \cos \alpha$.

Answer A!

2. $\sin \alpha$ = opposite side to α / hypotenuse = v_{0y} / v_0. $\Rightarrow v_{0y} = v_0 \times \sin \alpha$.

HELP: v_{0y} = Initial vertical velocity in m/s. v_0 = Projection speed in m/s.

α = Projection angle with respect to the horizontal.

Answer C!

4. If acceleration is a constant, we can use the kinematics equations.

$$v_{fy} = v_{0y} + (a_y \times t) \qquad\qquad \textbf{(1)}$$
$$v_{fy}^{\,2} = v_{0y}^{\,2} + (2 \times a_y \times \Delta y) \qquad\qquad \textbf{(2)}$$
$$\Delta y = (v_{0y} \times t) + (\tfrac{1}{2} \times a_y \times t^2) \qquad \textbf{(3)}$$

HELP: v_{0y} = Initial vertical *velocity* in m/s. v_{fy} = Final vertical *velocity* in m/s.

v_0 = Projection speed in m/s. α = Projection angle *with respect to the horizontal*.

a_y = acceleration along the vertical in m/s/s. t = time interval in s. Δy = Displacement along the vertical in m between the initial and final y positions.

Now we need to apply these equations with what we know. We are asked to determine the final speed along *y* after a time *t*.

We know v_0 and α. That means we know the initial vertical velocity component v_{0y} in terms of projection speed v_0 and projection angle α. We are given *t*. Acceleration along the vertical is just the acceleration due to gravity. If we assume "up" as positive, then:

$a_y = -g$. Substitute everything we know into the proper equation and solve.

Use equation # **1** with $v_{0y} = v_0 \times \sin \alpha$, $a_y = -g$ in $v_{fy} = v_{0y} + a_y \times t$:

$\Rightarrow v_{fy} = (v_0 \times \sin \alpha) + [(-g) \times t]$.

$\Rightarrow v_{fy} = (v_0 \times \sin \alpha) - (g \times t)$.

Answer D!

5.   If there is no air resistance, the horizontal acceleration is zero. There is nothing that is pulling or pushing that will change the speed along the horizontal. Along the vertical, we have the earth pulling on us.
The velocity along the horizontal will never change if horizontal acceleration is zero.
Answer E!

7.   If a vector is given, we can determine the horizontal and vertical components if we know the angle and the magnitude of the velocity vector.
If instead we know the horizontal x and vertical y components, we can find the magnitude of the velocity and the angle at which that vector acts. I like that.
The magnitude of a vector whose x– (horizontal) and y– (vertical) components are known is given by:
$|v|_{\text{MAGNITUDE}} = (v_{fx}^2 + v_{fy}^2)^{\frac{1}{2}}$.
(Not asked: The direction of the vector θ with respect to the horizontal: $tan\ \theta = v_{fy} / v_{fx}$.)
Answer A!

8.   If the x and y components are known, the heading of a vector is given by: $tan\ \theta = v_{fy} / v_{fx}$.
Kinematics equation **(1)** of solution to problem **4**, $v_{fy} = v_{0y} + (a_y \times t)$ becomes:
HELP: θ = Angle of the velocity vector with respect to the horizontal.
v_{fy} = The final vertical velocity in m/s.
v_{fx} = The final horizontal velocity in m/s = The initial horizontal velocity = $(v_0 \times cos\ \alpha)$.
v_{0y} = Initial vertical velocity = $v_0 \times sin\ \alpha$. v_0 = Projection speed in m/s.
α = The projection angle of the initial velocity vector with respect to the horizontal.
a_y = Acceleration along the vertical in m/s/s = $-g$. t = The time interval in s.

$v_{fy} = (v_0 \times sin\ \alpha) + (-g \times t)$. See also solution to problem **2**.
$v_{fx} = v_{0x} = v_0 \times cos\ \alpha$.
Substitute for v_{fx} and v_{fy} in $tan\ \theta = v_{fy} / v_{fx}$ and we get:
$tan\ \theta = [(v_0 \times sin\ \alpha) - (g \times t)] / (v_0 \times cos\ \alpha)$.
Answer C!

10.   Of the three equations of kinematics applied in the vertical direction, only two of them contain the vertical displacement between the initial and final vertical positions, Δy:
$v_{fy}^2 = v_{0y}^2 + (2 \times a_y \times \Delta y)$ **(2)**
$\Delta y = (v_{0y} \times t) + [(\frac{1}{2}) \times a_y \times t^2]$ **(3)**
HELP: Initial velocity along the vertical = $v_{0y} = (v_0 \times sin\ \alpha)$, and acceleration along the vertical
$a_y = -g$ in equation **(3)**:
$\Delta y = [(v_0 \times sin\ \alpha) \times t] + [\frac{1}{2} \times (-g) \times t^2]$.
$\Delta y = [(v_0 \times sin\ \alpha) \times t] - [\frac{1}{2} \times g \times t^2]$.
(By the way, we could have used equation **(2)** but the algebra would have been more complicated.)
Answer D!

11. From the third equation of kinematics for displacement between the initial and final y – positions,
$\Delta y = (v_{0y} \times t) + [(\frac{1}{2}) \times a_y \times t^2]$, we get:
$\Delta y = [(v_0 \times sin\ \alpha) \times t] - [\frac{1}{2} \times g \times t^2]$.
HELP: Initial velocity along the vertical = $v_{0y} = (v_0 \times sin\ \alpha)$, and acceleration along the vertical $a_y = -g$.

    1.   The right side of the equation for vertical displacement
        $\{\Delta y = [(v_0 \times sin\ \alpha) \times t] - [\frac{1}{2} \times g \times t^2]\}$ is positive when:
        $[(v_0 \times sin\ \alpha) \times t]$ is greater than $[\frac{1}{2} \times g \times t^2]$. That is, the final vertical position is higher than the
        initial vertical position when the magnitude of $[(v_0 \times sin\ \alpha) \times t]$ is greater than the magnitude of
        $[\frac{1}{2} \times g \times t^2]$.

    2.   When vertical displacement Δy is zero, the right side is equal to zero.
        $\Delta y = 0 = [(v_0 \times sin\ \alpha) \times t] - [\frac{1}{2} \times g \times t^2]$
        $\Rightarrow [(v_0 \times sin\ \alpha) \times t] = \frac{1}{2} \times g \times t^2$.
        Cancel the time t through out. $\Rightarrow (v_0 \times sin\ \alpha) = \frac{1}{2} \times g \times t$.

    3.   The vertical displacement is negative (the final vertical position is below the initial vertical position)
        when the magnitude of $[(v_0 \times sin\ \alpha) \times t]$ is less than the magnitude of $[\frac{1}{2} \times g \times t^2]$.

Answer B!

13. The horizontal displacement $\Delta x = v_{0x} \times t$.
Or $\Delta x = (v_0 \times cos\ \alpha) \times t$.
From the third equation of kinematics, we get: $\Delta y = [(v_0 \times sin\ \alpha) \times t] - [\frac{1}{2} \times g \times t^2]$.
When the starting and finishing points are at the same level, the vertical displacement $\Delta y = 0$ meters. $\Rightarrow 0 = \Delta y = [(v_0 \times sin\ \alpha) \times t] - [\frac{1}{2} \times g \times t^2]$.
Divide both sides by time t, and we get: $0 = (v_0 \times sin\ \alpha) - (\frac{1}{2} \times g \times t)$.
Solve for $t = (2 \times v_0 \times sin\ \alpha) / g$.
Substitute for t in the equation: $\Delta x = (v_0 \times cos\ \alpha) \times t$, and we get:
$\Delta x = (v_0 \times cos\ \alpha)_{\text{HORIZONTAL VELOCITY IN m/s}} \times [(2 \times v_0 \times sin\ \alpha) / g]_{\text{TIME IN s}}$.
[Note: The horizontal displacement when $\Delta y = 0$ m, is called the range R.]
Answer A!

14. Horizontal range is defined as the value of the horizontal displacement Δx when the evrtical displacement Δy
= 0 m. If the initial point from which you project is
$x = 0$ m, $y = 0$ m, then you can define range as the value of horizontal displacement when $\Delta y = 0$ m as $x_f - x_i$
$= x - 0 = x$ m $= R$.
Answer C!

16. Even though the velocity along the vertical is zero at the maximum height, the acceleration along the vertical
is not. Earth pulls on the projectile in the vertically down direction for all points of the projectile trajectory.
Acceleration at the maximum height is the acceleration due to gravity. (The horizontal acceleration is always
zero in the absence of horizontal forces.)
Answer A!

17. Use equation **2**, defining maximum height as the "final" position for this problem.
    HELP: Initial vertical position $y_i = 0$ m. Final vertical position $y_f = h$ m. The speed along the vertical is zero at the maximum height. $v_{fy} = 0$ m/s.
    The acceleration along the vertical $a_y = -g = -10$ m/s/s.
    $v_{fy}^2 = v_{0y}^2 + (2 \times a_y \times \Delta y)$.
    $0 = (v_{0y} \times sin\ \alpha)^2 + [2 \times (-g) \times (h - 0)]$.
    Here h is the vertical height from the initial position.
    Or $h = (v_0 \times sin\ \alpha)^2 / 2g = (v_0^2 \times sin^2\ \alpha) / 2g$.
    Answer D!

19. If Δy is zero, the position of maximum height h is exactly at the middle or when $x = R / 2$ m.
    Answer D!

20. Use equation **1**. $v_{fy} = v_{0y} + (a_y \times t)$.
    HELP: At the maximum height the vertical velocity $v_{fy} = 0$. The acceleration along the vertical $a_y = -g = -10$ m/s/s. t is the time to reach the maximum height from the initial launch point. The initial vertical velocity $v_{0y} = v_0 \times sin\ \alpha$ where v_0 is the projection speed and α is the projection angle in degrees with respect to the horizontal.
    $\Rightarrow 0 = (v_0 \times sin\ \alpha) + [(-g) \times t]$. $\Rightarrow t = (v_0 \times sin\ \alpha) / g$.
    Answer A!

22. Vertical displacement when the ball touches the ground is: $\Delta y = y_f - y_0 = 0 - 0 = 0$ meters.
    Along x, $\Delta x = v_{0x} \times t = (v_0 \times cos\ \alpha) \times t$.
    We need to solve for time t to reach the ground from the initial launch point.
    The final velocity along y is the same as initial velocity along y in magnitude but opposite in sign as the object is coming down. $v_{fy} = -v_{0y} = -v_0 \times sin\ \alpha$, just before the object hits the ground.
    The first equation of kinematics; $v_{fy} = v_{0y} + a_y \times t$ becomes:
    $(-v_0 \times sin\ \alpha) = (v_0 \times sin\ \alpha) + [(-g) \times t]$.
    $\Rightarrow t = (2 \times v_0 \times sin\ \alpha) / g$.
    [Note that this time t is twice much as the result of problem **20**. You should not be surprised. Time to hit the ground from the launching point is twice the time it takes to reach the maximum height if the initial vertical launching position for the projectile is at the same level as the landing position.]
    This gives us the horizontal displacement: $\Delta x = (v_0 \times cos\ \alpha) \times [2 \times v_0 \times sin\ \alpha / g]$
    $\Rightarrow \Delta x = v_0^2 \times (2 \times sin\ \alpha \times cos\ \alpha) / g$.
    Use a small trick: $sin\ 2\alpha = 2\ sin\ \alpha \times cos\ \alpha$.
    $\Rightarrow sin\ \alpha \times cos\ \alpha = sin\ 2\alpha / 2$. Substitute back and we get:
    $\Delta x = (v_0^2 \times sin\ 2\alpha) / g$.
    Vertical displacement = 0 m. Horizontal displacement, $\Delta x = v_0^2 \times sin\ 2\alpha / g$.
    Answer A!

23. Let me ask you the question another way. At what angle should you throw to make a ball go the maximum distance? Solution to problem **22** gives you the result for the range as:

$\Delta x = (v_0{}^2 \times sin\ 2\alpha)\ /\ g$. This is maximum (for a fixed projection speed v_0), when $sin\ 2\alpha$ is maximum or when $sin\ 2\alpha = 1$ because sin of any angle cannot exceed 1. Note that $sin\ 90 = 1$. The maximum value of sin θ does not exceed one. This ($sin\ 2\alpha = 1$) is possible only when $2\alpha = 90$ or when $sin\ 90 = 1$. $\Rightarrow\ \alpha = 90\ /\ 2$ or 45 degrees.
   Answer C!

25. We are looking for the position of the projectile.
   From the third equation of motion: $\Delta y = (v_{0y} \times t) + [(\tfrac{1}{2}) \times a_y \times t^2]$.
   We see that the ball does not follow a straight line because of the square term involving time.
   If $\Delta y = v_{0y} \times t$, then it would have been a straight line. The square term for time makes it a parabola.
   Answer C!

26. Acceleration along the horizontal is zero, so the final speed along the horizontal will not change:
   $v_{fx} = v_{0x} = v_0 \times cos\ \alpha$.
   Initially the speed along the vertical is $v_{0y} = v_0 \times sin\ \alpha$. At that instant, it is on the way up, so it is positive. Just before the projectile hits the ground, it is on the way down, and so the velocity vector along y should be negative. Use the second equation of motion: $v_{fy}{}^2 = v_{0y}{}^2 + (2 \times a_y \times \Delta y)$.
   $\Delta y = 0$.
   $\Rightarrow\ v_{fy}{}^2 = (v_0 \times sin\ \alpha)^2 + 0. \Rightarrow\ v_{fy}{}^2 = (v_0 \times sin\ \alpha)^2$.
   You might be tempted to say $v_{fy} = v_0 \times sin\ \alpha$. But the ball is on the way down just before it hits the ground. You should go with the *negative* root; $v_{fy} = -v_0 \times sin\ \alpha$.
   Answer C!

28. $\Delta y = (v_{0y} \times t) + (\tfrac{1}{2} \times a_y \times t^2)$. Because of t^2 term, if we plot Δy and time, we get a parabola that is symmetric. Remember that Δy is zero when the ball touches the ground.
   Answer C!

29. The velocity along the horizontal should not change at all as acceleration along the horizontal is zero. There is no net force along the horizontal. Horizontal velocity-time graph is a straight line that is parallel to the x axis. Time is plotted on the x axis. Horizontal velocity is a constant in time just before the projectile strikes the ground.
   Answer B!

31. Along x, acceleration is zero. There are no forces that we know of along the horizontal.
   Answer E!

32. Close to the surface of the earth, acceleration due to gravity is a constant. Along y, acceleration is a constant $= -g = -10$ m/s/s.
   Answer B!

CHAPTER 2, Solution Page 5

34.   Maximum range is attained when $\theta = 45$. No, it is not zero degrees with respect to the horizontal! The range
       equation: Horizontal displacement $\Delta x = (v_0^2 \times sin\, 2\alpha) \,/\, g$ is a maximum when $\theta = 45$. [Please look at
       solutions to problems **22** and **23**.]
       Range is zero when $\theta = 90$ or projected vertically up.
       Maximum height is obtained when $\theta = 90$ or when thrown vertically up.
       [When thrown at an angle, only a component of v_0 comes into the game.]
       Answer C!

35.   We want to negate the pull of the gravity. For this reason, we aim higher at an angle above the horizontal.
       Answer B!

37.   Anything that is projected along the horizontal will have a bigger horizontal speed than what is projected at
       an angle with the same speed v_0.
       Horizontal component for **Figure A** $= v_0 = 10$ m/s.
       Horizontal component for **Figure B** $= 0$ m/s.
       Horizontal component for **Figure C** $= 0$ m/s.
       Horizontal component for **Figure D** $= v_0 \times cos\, 30 = 10\; cos\, 30 = 8.7$ m/s. (Less than v_0).
       Horizontal component for **Figure E** $= v_0 \times cos\, 30 = 10\; cos\, 30 = 8.7$ m/s. (Less than v_0).
       Answer A!

38.   For **Figure A**, $v_{0y} = 0$ m/s.
       It makes sense. Projection is horizontal. $\Rightarrow$ Vertical component is zero.
       For **Figure B**, $v_{0y} = v_0 = 10$ m/s. Projection is strictly along the vertical.
       For **Figure C**, $v_{0y} = -v_0 = -10$ m/s. It is thrown down. That is why velocity is negative.
       For **Figure D**, $v_{0y} = v_0 \times sin\, 30 = 10 \times sin\, 30 = 5$ m/s.
       For **Figure E**, $v_{0y} = -v_0 \times sin\, 30 = -10 \times sin\, 30 = -5$ m/s.
       Definitely figures **B** and **C** have the largest vertical speed (magnitude) at the start.
       Answer B!

40.   Let us calculate the time taken by the object to hit the ground in each case.
       In all the five cases, the vertical displacement between the initial and final positions is
       -20 m $= \Delta y$. If you know that, you are ahead of others! Good.
       Figure A: Use the third equation of kinematics for constant acceleration along y.
       $\Delta y = (v_{0y} \times t) + (\tfrac{1}{2} \times a_y \times t^2)$.
       Substitute: $\Delta y = -20$ m, $v_{0y} = 0$ m/s, and $a_y = -g = -10$ m/s^2.
       $\Rightarrow\; -20 = (0 \times t) + [\tfrac{1}{2}\,(-10)\, t^2]$. Solve for $t = 2$ s.

       Figure B: Substitute $\Delta y = -20$ m, $v_{0y} = 10$ m/s, and $a_y = -g = -10$ m/s^2.
       $\Delta y = (v_{0y} \times t) + (\tfrac{1}{2} \times a_y \times t^2)$.
       $-20 = (10 \times t) + (\tfrac{1}{2} \times -10 \times t^2)$.
       This is a quadratic equation in time. Solve for positive roots $t = 3.24$ s.

CHAPTER 2, S o l u t i o n   P a g e   **6**

Figure C: Substitute $\Delta y = -20$ m, $v_{0y} = -10$ m/s, and $a_y = -g = -10$ m/s^2.
$\Delta y = (v_{0y} \times t) + (\frac{1}{2} \times a_y \times t^2)$.
$-20 = (-10 \times t) + (\frac{1}{2} \times a_y \times t^2)$. This is a quadratic equation in time. Solve for $t = 1.24$ s.
Figure D: Substitute $\Delta y = -20$ m, $v_{0y} = (10 \times \sin 30)$ m/s, and $a_y = -g = -10$ m/s^2.
$\Delta y = (v_{0y} \times t) + (\frac{1}{2} \times a_y \times t^2)$.
$-20 = [(10 \times \sin 30) \times t] + (\frac{1}{2} \times a_y \times t^2)$.
This is a quadratic equation in time. Solve for positive roots $t = 2.56$ s.

Figure E: Substitute $\Delta y = -20$ m, $v_{0y} = (-10 \times \sin 30)$ m/s, and $a_y = -g = -10$ m/s^2.
$-20 = [(-10 \times \sin 30) \times t] + (\frac{1}{2} \times a_y \times t^2)$.
This is a quadratic equation in time. Solve for positive roots $t = 1.56$ s.
Answer E!

41.   No doubts about this one: The one that is thrown vertically down.
Answer C!

43.   Let us find the vertical velocity by using the first equation of kinematics:
$v_{fy} = v_{0y} + (a_y \times t)$.
Figure A: $v_{0y} = 0$ m/s, $a_y = -10$ m/s^2. Time is calculated in problem **40**, $t = 2$ s.
$v_{fy} = 0 + [(-10) \times t]$
$\Rightarrow v_{fy} = (-10) \times 2 = -20$ m/s.

Figure B: $v_{0y} = 10$ m/s, $a_y = -10$ m/s^2. Time is calculated in problem **40**, $t = 3.24$ s.
$v_{fy} = 10 + [(-10) \times t]$.
$\Rightarrow v_{fy} = 10 + [(-10) \times 3.24] = -22.4$ m/s.

Figure C: $v_{0y} = -10$ m/s, $a_y = -10$ m/s^2. Time is calculated in problem **40**, $t = 1.24$ s.
$v_{fy} = (-10) + [(-10) \times t]$.
$\Rightarrow v_{fy} = (-10) + [(-10) \times 1.24] = -22.4$ m/s.

Figure D: $v_{0y} = (10 \times \sin 30) = 5$ m/s, $a_y = -10$ m/s^2. Time is calculated in problem **40**,
$t = 2.56$ s.
$v_{fy} = 5 + [(-10) \times t]$.
$\Rightarrow v_{fy} = 5 + [(-10) \times 2.56] = -20.6$ m/s.

Figure E: $v_{0y} = -10 \times \sin 30 = -5$ m/s, $a_y = -10$ m/s^2.
Time is calculated in problem **40**, $t = 1.56$ s.
$v_{fy} = (-5) + [(-10) \times t]$.
$\Rightarrow v_{fy} = (-5) + [(-10) \times 1.56] = -20.6$ m/s.
Answer A!

44.   With respect to the horizontal x, the three equations of kinematics with zero acceleration reduce to:

$\Delta x = v_{0x} \times t$. Note that we already calculated the time in **40**.

Δx for **Figure A** = $(10 \text{ m/s}) \times (2 \text{ s}) = 20 \text{ m}$.

Δx for **Figure B** = $(0) \times (3.24 \text{ s}) = 0 \text{ m}$.

Δx for **Figure C** = $(0) \times (1.24 \text{ s}) = 0 \text{ m}$.

Δx for **Figure D** = $(10 \cos 30) \times (1.56 \text{ s}) = 22.3 \text{ m}$.

Δx for **Figure E** = $(10 \cos 30) \times (2.56 \text{ s}) = 13.6 \text{ m}$.

Answer D!

46.   Use $\tan \theta = v_{fy} / v_{fx}$ to get the heading. We already calculated v_{fy} in **43**. The horizontal component of the velocity was calculated in **37**.

Figure A: $v_{fy} = -20 \text{ m/s}$, $v_{fx} = v_{0x} = 10 \text{ m/s}$.

         $\tan \theta = 20 / 10$.

         Solve for $\theta = 63.4°$ with respect to the horizontal.

Figure B: $v_{fy} = -22.4 \text{ m/s}$, $v_{fx} = v_{0x} = 0 \text{ m/s}$.

         $\tan \theta = 22.4 / 0$.

         $\theta = 90°$ with respect to the horizontal x axis.

Figure C: $v_{fy} = -22.4 \text{ m/s}$, $v_{fx} = v_{0x} = 0 \text{ m/s}$.

         $\tan \theta = 22.4 / 0$.

         $\theta = 90°$ with respect to x axis.

Figure D: $v_{fy} = -20.6 \text{ m/s}$, $v_{fx} = 8.7 \text{ m/s}$.

         $\tan \theta = 20.6 / 8.7$.

         $\theta = 67°$ with respect to the horizontal x axis.

Figure E: $v_{fy} = -20.6 \text{ m/s}$, $v_{fx} = 8.7 \text{ m/s}$.

         $\tan \theta = 20.6 / 8.7$.

         $\theta = 67°$ with respect to the horizontal x axis.

Answer E!

47.   Use the third equation of kinematics along y: $\Delta y = (v_{0y} \times t) + (\frac{1}{2} \times a_y \times t^2)$.

(A)   Ball dropped from rest:

         $v_{0y} = 0 \text{ m/s}$, $a_y = -10 \text{ m/s}^2$.

         $\Delta y = 0 + [\frac{1}{2} \times (-10) \times t_1^2]$.

(B)   Ball projected horizontally:

         $v_{0y} = 0 \text{ m/s}$. The third equation of kinematics becomes:

         $\Delta y = 0 + [\frac{1}{2} \times (-10) \times t_2^2]$.

Since the two balls are dropped and projected from the same height, Δy will be the same for both the balls. That means they will have to reach the ground at the same time if there is no air resistance.

Answer B!

49.   From problem **23** horizontal range, $R = (v_0^2 \, sin \, 2\alpha) \, / \, g$.

⇒ $R_{max} = v_0^2 \, / \, g$ is the maximum ($= R_{max}$), This occurs when $\alpha = 45$.

Let us call $R_{max} = v_0^2 \, / \, g$ as equation **(1)**.

At the maximum height, $v_{fy} = 0$. Use the second equation of kinematics for constant acceleration.

$v_{fy}^2 = v_{0y}^2 + (2 \times a_y \times \Delta y)$.

$0 = (v_0 \times sin \, 45)^2 + [2 \times (-g) \times h]$.

Note that $v_{0y} = v_0 \times sin \, 45$!

⇒ $h = (v_0 \times sin \, 45)^2 \, / \, 2g$.

⇒ $h = (v_0^2 \times sin^2 \, 45) \, / \, 2g$.

Substitute for $sin \, 45$. ⇒ $h = v_0^2 \, / \, 4g$ **(2)**.

From **(1)** and **(2)** we see that: $h = R_{max} \, / \, 4$.

Answer D!

50.   At the maximum height, $v_{fy} = 0$. Use the second equation of kinematics for constant acceleration.

$v_{fy}^2 = v_{0y}^2 + (2 \times a_y \times \Delta y)$.

⇒ $0 = [(v_0 \times sin \, \alpha)^2] + [2 \times (-g) \times h]$.

⇒ $h = (v_0 \times sin \, \alpha)^2 \, / \, (2 \times g)$.   **(1)**

From problem **22**, horizontal range $R = [v_0^2 \times (2 \times sin \, \alpha \times cos \, \alpha)] \, / \, g$. **(2)**

Divide **(1)** by **(2)**. This means that $h \, / \, R = sin \, \alpha \, / \, (4 \times cos \, \alpha)$.

Use $tan \, \alpha = sin \, \alpha \, / \, cos \, \alpha$.

This gives us $h \, / \, R = tan \, \alpha \, / \, 4$.

Answer B!

52.   The component of g parallel to the inclined plane $= g \times sin \, \alpha$ where α is now the angle of the inclined plane.
Answer C!

53.   The object is traveling down the inclined plane, so we expect speed to increase because acceleration and velocity are in the same direction.

We can use the equations of kinematics.

Take x as parallel to the inclined plane. (This is something that is allowed. It will not mess up your inclined plane stuff.)

$v_{0x} = 2$ m/s. $a_x = g \, sin \, 30 = 10 \times (0.5) = 5$ m/s^2. $t = 2$ s. $v_{fx} = ?$

Use the first equation of kinematics for constant acceleration. $v_f = v_0 + (a \times t)$.

$v_f = 2 + (5 \times 2) = 12$ m/s.

Answer C!

55.   When the object comes to a stop, its final speed is zero.

$v_{0x} = 30$ m/s, $v_{fx} = 0$ m/s, $a_x = g \times sin \, \alpha = 10 \times sin \, 30 = 5$ m/s^2.

While the object climbs up on the inclined plane, direction of displacement (or for that matter velocity) and acceleration are in the opposite directions. $a_x = -5$ m/s^2 (it is decelerating).

Use the second equation of kinematics. $v_{fx}^2 = v_0^2 + (2 \times a_x \times \Delta x)$.

$0 = (30 \text{ m/s})^2 + [2\,(-5) \times \Delta x]$. Solve for $\Delta x = 90$ m.
Answer D!

56. Once the object falls off, the only acceleration is along the vertical due to the earth's pull. $a_y = -10$ m/s per second.
Answer C!

58. Yes... When you drop a ball, displacement and acceleration are both in the vertically down direction.
Answer A!

59. Yes. When you throw a ball in the vertically up direction, the displacement is vertically up, but the acceleration due to gravity is acting vertically down.
Answer A!

61. Yes. Again when you throw a ball in the up direction, velocity is along the positive y direction if we define the y axis to be vertical, but acceleration is along the $-y$ direction.
Answer A!

62. "I saw that (...chalk dust move downward...)." I am quoting my son, Alex. He says Hi to you all.
Answer A!

64. Answer B!

65. When m_1 moves by 10 m, m_2 moves only by 5 m. $\Rightarrow$ Acceleration of m_2 is half the acceleration of m_1.
Answer B!

67. $\Delta x = v_{0x} \times t = (400 \text{ m/s}) \times t$ **(1)**
From the third equation of kinematics for constant acceleration, with $v_{0y} = 0$ m/s, becomes:
$\Delta y = \frac{1}{2} \times a_y \times t^2$.
$-20 = \frac{1}{2}(-10)\,t^2$.
$t = 2$ seconds.   **(2)**
$\Rightarrow \quad \Delta x = (400 \text{ m/s}) \times (2 \text{ s}) = 800$ m.
Answer D!

68. Since A and B are at the same vertical level, velocity along the vertical at A will be in the up direction, while vertical velocity at B will be in the down direction. They are equal in magnitude.
Answer A!

70. Time to rise = Time to fall back.
Use the first equation of kinematics.

On the way up:

$v_{fy} = v_{0y} + (a_y \times t)$.

At the maximum height, $v_{fy} = 0$.

$\Rightarrow 0 = 10 + [(-10) \times t]$.

$\Rightarrow t = 1$ s.

It takes 1 s to go up.

$\Rightarrow$ It will take exactly 1 s to come down in the absence of air resistance.

Total time = 1 s + 1 s = 2 s.

Answer B!

71.   Even though the clown throws the dummy vertically up, both clown and dummy have the same horizontal speed of 10 m/s. So they both travel the same distance along the horizontal. (Clown will be able to catch the dummy!)

Answer C!

73.   The train is already moving at a speed of 30 m/s.

That means the tomato also is moving at 30 m/s just before it is thrown.

When the tomato itself is thrown at 20 m/s and 75 degrees angle with repect to the horizontal, the x components of the two velocities add.

$v_{tomato} = 30$ m/s + $(20 \times cos\ 75)$.

Answer A!

74. $\Delta y = (v_{0y} \times t) + (\frac{1}{2} \times a_y \times t^2)$.

(12 m – 3 m) = $(v_{0y} \times 3$ s$) + [\frac{1}{2}(-10$ m/s$^2) \times (3$ s$)^2]$.

$\Rightarrow 9 = 3v_{0y} - 45$.

$\Rightarrow 3v_{0y} = 54$.

$\Rightarrow v_{0y} = 18$ m/s

Answer D!

76. $v_{0x} = v_0\ cos\ \theta = 12$ m/s. Given in problem **76**.

$v_{0y} = v_0\ cos\ \theta = 18$ m/s. From solution **74**.

$\Rightarrow tan\ \theta = 18 / 12$ or $\theta = 56.3$ degrees with respect to the horizontal.

Answer E!

1.   Answer C!

2.   Answer D!

4.   Answer B!

5.   Answer A!

7.   Answer D!

8.   Answer E!

10.   Answer D!

11.   Answer D!

13.   Answer C!

14.   Answer C!

16.   Answer A!

17.   Answer D!

19.   Answer E!

20.   Answer A!

22.   Answer B!

23.   Answer E!

25.   Answer A!

26.   Answer A!

28.   Answer C!

29.   Answer B!

31.   Answer A!

32.   Answer A!

34.   Answer D!

35.   Answer C!

37.   Answer D!

38.   Answer A!

40.   Answer A!

41.   Answer A!

43.   Answer A!

44.   Answer A!

46.   Answer A!

47.   Answer B!

49.   Answer A!

50.   Answer C!

52.   Answer A!

53.   Answer C!

55.   Answer B!

56.   Answer A!

58.   Answer C!

59.   Answer A!

61.   If the train does not accelerate, the ball should come back into the thrower's hand, because the ball and the train have the same horizontal speed.
Answer B!

62. Net force along y: $\sum F_y = m \times a_y$.

Force from the air resistance acts vertically upwards. Force of weight acts vertically downward.

By common sense: $(m_1 \times g) - F_R = m_1 \times a_1$ and $(m_2 \times g) - F_R = m_2 \times a_2$.

By vector sum of components: $(-m_1 \times g) + F_R = -m_1 \times a_1$.

$(-m_2 \times g) + F_R = -m_2 \times a_2$.

$\Rightarrow a_1 = [(m_1 \times g) - F_R] / m_1$ and $a_2 = [(m_2 \times g) - F_R] / m_2$.

$a_1 = g - (F_R / m_1)$ and $a_2 = g - (F_R / m_2)$.

F_R / m_1 is smaller than F_R / m_2 as $m_1 > m_2$. That means acceleration of m_1 is larger than acceleration of m_2.

Answer B!

64. The elevator is accelerating up, meaning that the sum of all forces along positive y should be greater than the sum of all forces along negative y.

By common sense: $N - (m \times g) = m \times a$.

By vector sum of components: $N + (-m \times g) = (+m \times a)$.

$\Rightarrow N - m \times g = m \times a$. $\Rightarrow N = m \times (g + a)$.

$\Rightarrow N = 60 \times (10 + 2)$. $\Rightarrow N = 720\,\text{N}$.

Answer E!

65. When the elevator is descending at a constant acceleration of $2\,\text{m/s}^2$, the sum of all the forces along $-y$ should be greater than the sum of all the forces along $+y$.

By common sense: $(m \times g) - N = m \times a$,

$N = (m \times g) - (m \times a) = m \times (g - a)$.

By vector sum of components: $N + (-m \times g) = -m \times a$.

$N = 60 \times (10 - 2)$. $\Rightarrow N = 480\,\text{N}$.

Answer A!

67. When the elevator is rising at uniform speed, its velocity is constant. If there is no change in the velocity vector, acceleration will be zero. That means the vector sum of all forces is zero.

$N + (-m \times g) = 0$ or $N = m \times g = 60 \times 10 = 600\,\text{N}$.

Answer B!

68. When the elevator is descending at constant speed, its velocity is a constant. This gives us change in velocity equal to zero and acceleration equal to zero. If acceleration is zero, the vector sum of all forces should be zero.

$N - (m \times g) = 0$ or $N = m \times g$. $\Rightarrow N = 60\,\text{kg} \times 10\,\text{m/s}^2 = 600\,\text{N}$.

Answer B!

70. When the elevator is rising at a constant acceleration of $3\,\text{m/s}^2$, the sum of all forces along $+y$ has to be greater than the sum of all forces along $-y$. The elevator accelerates upwards because there is a net force along $+y$.

By common sense: $T - (m \times g) = m \times a$.

By vector sum of components: $T + (- m \times g) = m \times a.$
$\Rightarrow T = (m \times g) + (m \times a) = m \times (g + a). \Rightarrow T = (0.1) \times (10 + 3)$
$\Rightarrow T = 1.3 \text{ N}.$

Answer B!

71.   When the elevator descends at a constant acceleration, net force is in the "down" direction.
By common sense: $(m \times g) - T = m \times a.$
By vector sum of components: $T + (- m \times g) = (- m \times a).$
$\Rightarrow T = (m \times g) - (m \times a). \Rightarrow T = m \times (g - a). \Rightarrow T = 0.1 \times (10 - 3).$
$\Rightarrow T = 0.7 \text{ N}.$

Answer A!

73.   If the elevator prepares to stop while descending, we get $T + (- m \times g) = m \times a.$
$\Rightarrow T = m \times (g + a).$
$\Rightarrow T = 0.1 \times (10 + 3). \Rightarrow T = 1.3 \text{ N}.$

Answer B!

74.   The elevator is accelerating up, meaning that the cable is also accelerating up.
<u>For the cable accelerating up:</u>
By common sense: $T_1 - [T_2 + (m_c \times g)] = m_c \times a$ **(1)**
By vector sum of components: $T_1 + (- T_2) + (- m_c \times g) = m_c \times a.$

<u>For the elevator accelerating up:</u>
By common sense: $T_2 - (m_e \times g) = m_e \times a.$ **(2)**
By vector sum of components: $T_2 + (- m_e \times g) = m_e \times a.$

Add the left sides of **(1)** and **(2)** and we get:
$T_1 - T_2 - (m_c \times g) + T_2 - (m_e \times g) = (m_c \times a) + (m_e \times a).$
$\Rightarrow T_1 - g \times (m_c + m_e) = a \times (m_c + m_e),$
$T_1 = a \times (m_c + m_e) + g \times (m_c + m_e),$
$T_1 = (a + g) \times (m_c + m_e),$
or $T_1 = (m_e + m_c) \times (g + a).$

Answer C!

76.   Because $m_2 > m_1$, m_2 accelerates down while m_1 accelerates up.
<u>For mass m_1:</u>
By common sense: $T - (m_1 \times g) = m_1 \times a$ **(1)**
By vector sum of components: $T + (- m_1 \times g) = m_1 \times a.$

<u>For mass m_2:</u>
By common sense: $(m_2 \times g) - T = m_2 \times a$ **(2)**

By vector sum of components: $T + (- m_2 \times g) = - m_2 \times a$.

Adding equations (1) and (2) gives us:

$$T - (m_1 \times g) + (m_2 \times g) - T = (m_1 \times a) + (m_2 \times a),$$
$$(m_2 - m_1) \times g = (m_2 + m_1) \times a. \Rightarrow a = [(m_2 - m_1) / (m_1 + m_2)] \times g,$$
$$a = [(0.25 - 0.20) / (0.25 + 0.20)] \times 10,$$
$$a = 1.1 \, \text{m} / \text{s}^2.$$

Substitute the value of acceleration into equation (1) or (2) and solve for T.
Let us take equation (1).

$$T - (m_1 \times g) = m_1 \times a.$$
$$\Rightarrow T = (m_1 \times a) + (m_1 \times g).$$
$$T = m_1 \times (a + g). \Rightarrow T = 0.2 \times (1.1 + 10). \Rightarrow T = 2.22 \, \text{N}.$$

Answer A!

77. m_1 accelerates to the right. Since there is only one force along x, we have:

$$T = m_1 \times a. \qquad\qquad (1)$$

m_2 accelerates down:

By common sense: $(m_2 \times g) - T = m_2 \times a. \qquad (2)$

By vector sum of components: $T + (- m_2 \times g) = (- m_2 \times a)$.

By adding (1) and (2) we get:

$$T + (m_2 \times g) - T = (m_1 \times a) + (m_2 \times a),$$
$$m_2 \times g = (m_1 + m_2) \times a. \Rightarrow a = [m_2 / (m_1 + m_2)] \times g,$$
$$\Rightarrow a = [3 / (2 + 3)] \times 10. \Rightarrow a = 6 \, \text{m} / \text{s}^2.$$

Answer C!

79.   Draw the force body diagram for each of the masses.

Equation of motion for mass m_1:
Tension T_2 is to the right.

$$\sum F_x = m_1 \times a \Rightarrow T_2 = m_1 \times a. \qquad (1)$$
$$\sum F_y = 0 \Rightarrow N_1 = m_1 \times g \text{ (not needed)}$$

Equation of motion for mass m_2:
Tension T_1 is to the right. Tension T_2 is to the left.

$$\sum F_x = m_2 \times a \Rightarrow T_1 - T_2 = m_2 \times a. \qquad (2)$$
$$\sum F_y = 0 \Rightarrow N_2 = m_2 \times g \text{ (not needed)}$$

Equation of motion for m_3:

$$\sum F_x = m_3 \times a \Rightarrow F - T_1 = m_3 \times a \qquad (3)$$
$$\sum F_y = 0 \Rightarrow N_3 = m_3 \times g \text{ (not needed)}$$

Add the left sides of equations (1), (2), and (3) and set their sum equal to the sum of the right sides.

$$T_2 = m_1 \times a. \qquad\qquad (1)$$

$$T_1 - T_2 = m_2 \times a. \qquad\qquad (2)$$
$$F - T_1 = m_3 \times a. \qquad\qquad (3)$$

We have the sum:
$$T_2 + T_1 - T_2 + F - T_1 = (m_1 \times a) + (m_2 \times a) + (m_3 \times a),$$
$$F = (m_1 + m_2 + m_3) \times a.$$

Substitute the values of $F = 20$ N, $m_1 = 2$ kg, $m_2 = 3$ kg, and $m_3 = 5$ kg.
$$20 = (2 + 3 + 5) \times a. \Rightarrow 20 = 10 \times a. \Rightarrow a = 20 \, / \, 10 = 2 \, \text{m} \, / \, \text{s}^2.$$

To find T_1: Substitute the value of a into equation (3) and we get
$$F - T_1 = m_3 \times a. \Rightarrow 20 - T_1 = 5 \, \text{kg} \times 2 \, \text{m/s}^2. \Rightarrow T_1 = 10 \, \text{N}.$$

Answer A!

80. Substitute the value of a from problem **79** into equation (2) of the same problem **79**:
$$T_1 - T_2 = m_2 \times a.$$
We have: $T_2 = ?$, $T_1 = 10$ N (from problem **79**), $m_2 = 3$ kg, and $a = 2 \, \text{m} \, / \, \text{s}^2$.
$$10 - T_2 = 3 \times 2,$$
$$T_2 = 10 - 6 = 4 \, \text{N}.$$

Answer C!

82. Since m_1 and m_2 are connected, the acceleration of both the masses will be the same.
Answer B!

83. Since acceleration of $m_1 = 2 \, \text{m} \, / \, \text{s}^2$ (refer solution **79**), the speed increases at the rate of $2 \, \text{m} \, / \, \text{s}$ every second. If the speed was zero at the start, its speed after 1 second $= 0 + 2 = 2 \, \text{m} \, / \, \text{s}$.
Formula:
$$v_f = v_0 + (a \times t).$$
$$\Rightarrow v_f = 0 + (2 \, \text{m} \, / \, \text{s}^2 \times 1 \, \text{s}) = 2 \, \text{m} \, / \, \text{s}.$$

Answer D!

85. Draw the free body diagram for all the masses.
Equation for motion for mass m_1:
Tension T_1 is to the right.
$$\sum F_x = m_1 \times a \Rightarrow T_1 = m_1 \times a. \quad (1)$$
$$\sum F_y = 0. \Rightarrow N_1 = m_1 \times g \text{ (not needed)}.$$

Equation of motion for mass m_2:
Tension T_2 is to the right. Tension T_1 is to the left.
$$\sum F_x = m_2 \times a.$$
By common sense: $\qquad T_2 - T_1 = m_2 \times a. \qquad (2)$
By vector sum of components: $\quad T_2 + (-T_1) = m_2 \times a.$
$$\sum F_y = 0. \Rightarrow N_2 = m_2 \times g \text{ (not needed)}$$

Equation of motion for mass m_3:

$$\sum F_x = m_3 \times a$$

By common sense: $T_3 - T_2 = m_3 \times a.$ **(3)**

By vector sum: $T_3 + (-T_2) = m_3 \times a.$

Equation of motion for m_4:

$$\sum F_x = m_4 \times a.$$

By common sense: $(m_4 \times g) - T_3 = m_4 \times a$ **(4)**

By vector sum: $T_3 + (-m_4 \times g) = (-m_4 \times a)$

Solving for acceleration, a:

$$T_1 = m_1 \times a \qquad \textbf{(1)}$$
$$T_2 - T_1 = m_2 \times a \qquad \textbf{(2)}$$
$$T_3 - T_2 = m_3 \times a \qquad \textbf{(3)}$$
$$(m_4 \times g) - T_3 = m_4 \times a \qquad \textbf{(4)}$$

Add the above equations and we get:

$$T_1 + T_2 - T_1 + T_3 - T_2 + (m_4 \times g) - T_3 = (m_1 \times a) + (m_2 \times a) + (m_3 \times a) + (m_4 \times a),$$
$$m_4 \times g = (m_1 + m_2 + m_3 + m_4) \times a. \Rightarrow a = [m_4 / (m_1 + m_2 + m_3 + m_4)] \times g.$$
$$\Rightarrow a = [3 / (4 + 4 + 4 + 3)] \times 10. \Rightarrow a = 2 \, \text{m} / \text{s}^2.$$

Answer D!

86. $v_0 = 0$, $a = 2 \, \text{m} / \text{s}^2$, and $t = 2 \, \text{s}$.

Acceleration of the system is $2 \, \text{m} / \text{s}^2$. That means that the speed increases by $2 \, \text{m} / \text{s}$ every second.

Speed after 1 s is $2 \, \text{m} / \text{s}$ more than the speed at $t = 0 \, \text{s}$.

$$= 2 \, \text{m} / \text{s} + 0 = 2 \, \text{m} / \text{s}.$$

Speed after 2 s is $2 \, \text{m} / \text{s}$ more than the speed at $t = 1 \, \text{s}$:

$$= 2 \, \text{m} / \text{s} + 2 \, \text{m} / \text{s} = 4 \, \text{m} / \text{s}.$$

(Speed after one more second $= 4 + 2 = 6 \, \text{m} / \text{s}$.)

Answer C!

88.   In solution **85**, we solved for $a = 2 \, \text{m} / \text{s}^2$. Substitute the value of a into equation **(4)** of the same problem.

$$(m_4 \times g) - T_3 = m_4 \times a.$$

Substitute $m_4 = 3 \, \text{kg}$, $g = 10 \, \text{m} / \text{s}^2$, and $a = 2 \, \text{m} / \text{s}^2$. We get:

$$(3 \times 10) - T_3 = 3 \times 2. \Rightarrow T_3 = 24 \, \text{N}.$$

From equation **(3)** of problem **85**,

$$T_3 - T_2 = m_3 \times a.$$

Substitute $T_3 = 24 \, \text{N}$, $m_3 = 4 \, \text{kg}$, and $a = 2 \, \text{m} / \text{s}^2$. We get:

$$24 - T_2 = 4 \times 2.$$
$$T_2 = 24 - 8 = 16 \, \text{N}.$$

Answer C!

89. From problem **88**, $a = 2 \text{ m} / \text{s}^2$, $T_2 = 16 \text{ N}$, and from equation (2) of problem **85** we get:
$$T_2 - T_1 = m_2 \times a.$$
Substitute $T_2 = 16 \text{ N}$, $m_2 = 4 \text{ kg}$, and $a = 2 \text{ m} / \text{s}^2$. We get: $16 - T_1 = 4 \times 2$. $\Rightarrow T_1 = 16 - 8 = 8 \text{ N}$.
Answer B!

91. The component of the weight parallel to the inclined plane, $w_x = w \times \sin \alpha = m \times g \times \sin \alpha$.
Answer B!

92. The component of the weight perpendicular to the inclined plane, $w_y = w \times \cos \alpha$. $w_y = m \times g \times \cos \alpha$.
Answer C!

94. Draw the free body diagram for the two masses.
<u>Equation of motion for mass m_1:</u>
$$\sum F_x = m_1 \times a.$$
By common sense: $\quad T - (m_1 \times g \times \sin \alpha) = m_1 \times a.$ (1)
By vector sum of components: $T + (- m_1 \times g \times \sin \alpha) = m_1 \times a.$
$$\sum F_y = 0.$$
Normal Force, $N = m_1 \times g \times \cos \alpha.$

<u>Equation of motion for mass m_2:</u>
$$\sum F_y = m_2 \times a.$$
By common sense: $\quad (m_2 \times g) - T = m_2 \times a.$ (2)
By vector sum: $\quad T + (- m_2 \times g) = - m_2 \times a.$

Add equations (1) and (2):
$$T - (m_1 \times g \times \sin \alpha) + (m_2 \times g) - T = (m_1 \times a) + (m_2 \times a),$$
$$(m_2 \times g) - (m_1 \times g \times \sin \alpha) = (m_1 + m_2) \times a,$$
$$(3 \times 10) - (2 \times 10 \times \sin 30) = (2 + 3) \times a,$$
$$30 - 10 = 5 \times a,$$
$$a = 20 / 5 = 4 \text{ m} / \text{s}^2.$$
Note that our initial assumption was correct: it is accelerating up the inclined plane. If our assumption was wrong, we would have got a negative sign.
If $a = 4 \text{ m} / \text{s}^2$, then we substitute the value of a into equation (2):
$$(m_2 \times g) - T = m_2 \times a.$$
$$\Rightarrow 3 \times 10 - T = 3 \times 4. \Rightarrow T = 30 - 12 = 18 \text{ N}.$$
Answer C!

95. Draw the free body diagram for both the rope and the rectangular mass. Assume m_{ROPE} and m_{25} are the masses of the rope and 25-kg mass respectively.
<u>Equation of motion for the rope:</u>
Force F is to the right. Tension T_1 is to the left. It is accelerating to the right.

$$\sum F_x = m_{\text{ROPE}} \times a.$$
$$F - T_1 = m_{\text{ROPE}} \times a. \qquad \textbf{(1)}$$

Equation of motion for the 25 kg mass:

Tension T_1 is to the right.

$$T_1 = m_{25} \times a. \qquad \textbf{(2)}$$

Add the two equations:

$$F - T_1 + T_1 = (m_{\text{ROPE}} + m_{25}) \times a,$$
$$F = (2 \text{ kg} + 25 \text{ kg}) \times a,$$
$$50 \text{ N} = 27 \times a,$$
$$a = 50 / 27 = 1.85 \text{ m} / \text{s}^2.$$

Substitute the value of a into equation **2** and we get:

$$T_1 = m_{25} \times a.$$
$$T_1 = 25 \times 1.85,$$
$$T_1 = 46.3 \text{ N}.$$

Answer B!

1.   Answer A!

2.   The maximum static frictional force is $F_f = \mu N = \mu mg$. If the force is horizontal, there is no frictional force. If μ is coefficient of static friction, then $F_f \leq \mu mg$ (inequality!) But if mass is at rest on a horizontal table, force of friction, $F_f = 0$.
       Answer D!

4.   See problem 2. The maximum static frictional force is $F_f = \mu N = \mu \times [10 + w]$. But if mass is at rest on a horizontal table, $F_f = 0$.
       Answer D!

5.   The sliding body is in contact with the table. If it is not a bumpy ride, net force along the vertical will be zero. $\Rightarrow$ Net force along $y = 0$.
       Answer A!

7.   The only force acting against the motion is the frictional force. Frictional forces are always in the opposite direction to the direction of motion. $\Rightarrow$ Net force $= ma$. $\Rightarrow F_f = ma$. $\Rightarrow \mu N = ma$. $\Rightarrow \mu mg = ma$. $\Rightarrow a = \mu g$.
       Answer B!

8.   Normal force balances the weight if there are no other forces acting with components along y. $\Rightarrow N = mg$.
       Answer A!

10.   When friction is present, it acts in the direction opposite to the direction of displacement, to the left.
       $\Rightarrow$ Net force along the horizontal $= ma_x$. $\Rightarrow 15 \cos 30 - F_f = ma_x$. **(1)**
       Along y, we have: $N + (15 \times \sin 30) = mg$.
       Note that $15 \times \sin 30$ and normal force are acting in the same direction (vertically up) while mg acts vertically down. $\Rightarrow N = mg - (15 \sin 30)$.
       $\Rightarrow F_f = \mu N = \mu[mg - (15 \times \sin 30)]$. **(2)**
       Substitute for F_f in equation **(1)**.
       $\Rightarrow (15 \cos 30) - \mu[mg - (15 \sin 30)] = ma_x$.
       Answer E!

11.   15-N force has a positive y component. $\Rightarrow (15 \sin 30) + N = mg$.
       $\Rightarrow N = mg - (15 \sin 30)$.
       Answer B!

13.   Equations of kinematics can be used (only) if the acceleration is a constant.
       Answer A!

14.   Answer A!

16.   Answer B!

17.   The only force along the inclined plane that is pulling on the mass to go down is the component of the weight. *Net* force along $x = ma_x$. $\Rightarrow mg\ sin\ \theta = ma$.
Answer B!

19. $mg\ sin\ \theta$ acts to the left. F_f acts to the right. *Net* force along $x = ma_x$. $\Rightarrow mg\ sin\ \theta - F_f = ma_x$.
Answer B!

20. $F_f = \mu N$. [Oops: Be very careful when saying this statement. The correct statement is that $F_f \le \mu N$. If that is the case, the greatest possible frictional force is μN. For example, if a book is at rest on a horizontal table, $\mu \ne 0$, $N \ne 0$, but $F_f = 0$!]   For a mass kept on an inclined plane, $mg\ cos\ \theta$ acts in the opposite direction to the normal force. If the mass is in contact with the surface, net force along y should be zero.
$\Rightarrow$ The component of the weight along the perpendicular to the inclined plane and the normal force should balance each other.
$\Rightarrow N = mg\ cos\ \theta$. $\Rightarrow F_f = \mu mg\ cos\ \theta$.
Answer B!

22.   The 5-N force acts in the same direction as the normal force. $\Rightarrow 5 + N = mg\ cos\ \theta$.
Answer B!

23.   Force of friction $= F_f = \mu N = \mu(mg\ cos\ \theta - 5)$. Note that $5 + N = mg\ cos\ \theta$.
Answer C!

25.   Tension T acts along the inclined plane to the right. $m_1 g\ sin\ \theta$ acts along the inclined plane to the left.
Net force along $x = m_1 a_x$. $\Rightarrow (m_1 g\ sin\ \theta) - T = m_1 a$.
Answer B!

26.   For the mass m_2: Tension T acts to the left parallel to the inclined plane. $m_2 g\ sin\ \theta$ acts to the left in the same direction as T. Net force along $x = m_2 a_x$. $\Rightarrow (m_2 g\ sin\ \theta) + T = m_2 a$.
Answer C!

28.   For the mass m_1: Tension T acts to the right. $m_1 g\ sin\ \theta$ acts to the left. It is sliding down, so force of friction is to the right. Net force along $x = m_1 a_x$.
$\Rightarrow (m_1 g\ sin\ \theta) - T - F_f = m_1 a$.
$\Rightarrow m_1 g\ sin\ \theta - (T + F_f) = m_1 a$.
Answer D!

29. For the mass m_2: Tension T acts to the left. The component of the weight, $m_2g \sin \theta$, also acts to the left. The mass m_2 is sliding to the left, so force of friction is to the right. $\Rightarrow T + m_2g \sin \theta - F_f = m_2a$.
But, $F_f = \mu N_2 = \mu m_2g \cos \theta$. $\Rightarrow T + (m_2g \sin \theta) - (\mu m_2g \cos \theta) = m_2a$.
Answer C!

31. For the mass m_2: 25-N force acts to the right. Tension T acts to the left. The component of the weight acts to the left. Net force along $x = m_2a_x$. $\Rightarrow 25 - T - (m_2g \sin \theta) = m_2a$.
Answer C!

32. For the mass m_1: The normal force, $N_1 = m_1g \cos \theta$. For the mass m_2, the normal force, $N_2 = m_2g \cos \theta$.
Answer A!

34. For mass m_1: The component of the weight, $m_1g \sin \theta_1$, acts to the left. Tension T acts to the right. $m_1 \gg m_2$. Assume m_1 accelerates to the left.
Net force along $x = m_1a_x$. $\Rightarrow (m_1g \sin \theta_1) - T = m_1a$.
Answer B!

35. For the mass m_2: Tension acts to the "left." The component of the weight, $m_2g \sin \theta_2$, acts to the right. The mass accelerates to the left. Net force along $x = m_2a_x$.
$\Rightarrow T - (m_2g \sin \theta_2) = m_2a$.
Answer D!

37. For the mass m_2: Tension acts to the left. The component of the weight of gravity, $m_2g \sin \theta_2$ acts to the right. $F_f = \mu N = \mu m_2g \cos \theta_2$ acts to the right. Net force along $x = m_2a_x$.
$\Rightarrow T - (m_2g \sin \theta_2) - F_f = m_2a$.
$\Rightarrow T - (m_2g \sin \theta_2) - (\mu m_2g \cos \theta_2) = m_2a$.
Answer C!

38. For the mass m_1: Tension T_1 acts vertically upwards. Weight, m_1g acts vertically down. The mass is accelerating down. Net force along $y = m_1a_y$. $\Rightarrow (m_1g) - T_1 = m_1a$.
Answer D!

40. For the mass m_3: Tension T_2 acts to the left. Net force along $x = m_3a_x$. $\Rightarrow T_2 = m_3a$.
Answer A!

41. If there is no contact, the force exerted by the surface on the mass is zero.
Answer A!

43. Net force along $y = m_1a_y = 0$. $\Rightarrow N_3 = m_3g$.
Answer B!

44. <u>For the mass m_2:</u> Tension T_1 acts to the left. Tension T_2 acts to the right. The component of the weight, $m_2g \sin \theta$, acts to the right. The mass m_2 is getting displaced to the left and so force of friction is to the right. $F_f = \mu N_2 = \mu (m_2g \cos \theta)$. Net force along $x = m_2a_x$.
$\Rightarrow T_1 - T_2 - (m_2g \sin \theta) - (\mu m_2g \cos \theta) = m_2a$.
Answer D!

46. m_1 accelerates down. Net force along $y = m_1a_y$.
$\Rightarrow (m_1g) - T = m_1a$.
Answer A!

47. m_2 accelerates upwards. Net force along $y = m_2a_y$. $\Rightarrow T - m_2g = m_2a$.
Answer D!

49. $F = m_1a_1 = m_2a_2$. $\Rightarrow a_1 = F / m_1$ and $a_2 = F / m_2$. $\Rightarrow a_1 << a_2$, if $m_1 >> m_2$. It is very easy to accelerate a small car compared to a giant truck. A small car needs only very small force.
Answer B!

50. $m_1g \sin \theta = m_1a_1$. $\Rightarrow a_1 = g \sin \theta$. Similarly, $m_2g \sin \theta = m_2a_2$. $\Rightarrow a_2 = g \sin \theta$. $\Rightarrow a_1 = a_2$.
Answer C!

52. The trick is to get your "opposite" angles to the three forces right. Also, this formula works only when there are THREE forces in equilibrium.
Answer A!

53. Answer B!

55. Answer A!

56. Answer A!

58. Answer A!

59. Did you forget to include the component of the force of gravity along the inclined plane? I hope not. Tension T_1 is to the right. $m_1g \sin \theta_1$ is also to the right. Force of friction is to the left.
Net force along $x = m_1a_x = 0$.
$\Rightarrow T_1 + (m_1g \sin \theta_1) = \mu m_1g \cos \theta_1$.
Answer B!

61. Answer A!

62. Answer B!

CHAPTER 4, Solution Page 4

1. Note that it is not necessary to say that the vector points at the center point O. The defining facts of a vector are its magnitude and direction, not where it is attached.
    Answer A!

2. Linear speed v is always equal to $R \times \omega$ and points along the tangent at the point of interest.
    Answer B!

4. Toward O from midpoint of *AB*. The direction of the change in velocity is along the line joining the particle at the midpoint of *AB* and the center of the circle.
    Answer B!

5. $v = R \times \omega$. If v is in m/s, then this equation is valid only if ω is in radians/second.
    Answer C!

7. Whenever an object moves in a circle, there is an acceleration *toward* the center along the radius.
    Answer A!

8. $v = R \times \omega \Rightarrow v$ is a constant if ω is a constant. Change in *speed* along the tangent is then zero.
    Answer B!

10. The *net* force acting on the particle *toward* the center: $F_c = m \times v^2 / R$.
    Using $v = R \times \omega$, we get
    $F_c = (m \times R^2 \times \omega^2) / R = m \times R \times \omega^2$.
    Answer A!

11. Angular velocity is related to linear speed: $v = R\omega$. If v is a constant, then ω will be a constant if R is fixed.
    Answer A!

13. Whenever an object moves in a circle, the acceleration is towards the center. At point A, it is along $-x$. At point B, it is along $-y$. At C, it is along $+x$. At D, the acceleration is along $+y$.
    Answer B!

14. If there is no skidding, force of friction, $\mu N = m \times v^2 / R$. $N = mg \Rightarrow \mu mg = m \times v^2 / R$.
    $\Rightarrow \mu = v^2 / (R \times g)$. If μ is less than $v^2 / (R \times g)$, it will skid. For the car not to skid, we need μ to be greater than $v^2 / (R \times g)$.
    Answer A!

16. Net force is inward at the bottom. Tension acts toward the center in the vertically up direction while weight acts downward. $\Rightarrow T - (m \times g) = m \times v^2 / R$. Also, $v = R \times \omega$ becomes $v = l \times \omega$.
    $\Rightarrow T - (m \times g) = [m \times (l \times \omega)^2] / l$. $\Rightarrow T = (m \times g) + (m \times l \times \omega^2)$.
    Answer A!

CHAPTER 5, Solution Page **1**

17. At the top of the circle, we have both tension and weight acting vertically downward toward the center.

Thus, the *net* force toward the center is $(m \times v^2) / R = T + (m \times g)$.

We find that this equation can be rewritten as $T = [(m \times v^2) / R] - (m \times g)$.

Substituting $v = R \times \omega$, we get $T = (m \times v^2 / R) - (m \times g)$.

Since $R = l$, the equation becomes $T = (m \times l \times \omega^2) - (m \times g)$.

Answer C!

19. At the top, *net* force is towards the center.

$\Rightarrow T + (m \times g) = (m \times v^2) / R$.

$\Rightarrow T = [(m \times v^2) / R] - (m \times g)$.

Substitute $v = R \times \omega$. $\Rightarrow T = [m \times (R \times \omega)^2 / R] - (m \times g)$.

$\Rightarrow T = (m \times R \times \omega^2) - (m \times g)$.

When the tension is zero, the particle meets the condition for "fall."

For vertical fall, we expect $m \times g > m \times R \times \omega^2$, so $g > (R \times \omega^2)$.

Dividing both sides by R gives $(g / R) > \omega^2$. Now solve for ω..

We get $\omega^2 < (g / R)$ or

$\omega < (g / R)^{1/2}$. Since $R = l$, $\omega < (g / l)^{1/2}$.

Answer A!

20. The time taken is: $t = distance\ from\ A\ to\ B / v$.

$\Rightarrow t =$ half the circumference $/ v = [(2 \times \pi \times R) / 2)] / v$.

$t_{AB} = (\pi \times R) / v$.

Answer A!

22. The average tangential acceleration between point A and point B is zero because the particle is moving at a constant speed.

Answer B!

23. At point A, radial acceleration is to the right, toward the center. At point B, radial acceleration is to the left, toward the center. Average radial acceleration is zero because they are equal in magnitude but opposite in direction.

Answer B!

25. Radial velocity is zero, but tangential velocity is not.

Answer B!

26. Since tangential velocity is not a constant, we have tangential acceleration. [Note: Whenever an object moves in a circle, we always have a radial acceleration.]

Answer C!

28.   Before the paint is ready to fall off (spill out), the forces acting on the paint are the weight acting vertically down and the normal force also acting vertically down. At the top, net force is inward toward the center of the circle. This makes the weight and the normal force added providing the centripetal acceleration $[= (m \times v^2) / R]$.

$\Rightarrow N + mg = (m \times v^2) / R.$ **(1)**.

When the paint is ready to fall off (spill out): $N = 0$ (paint is barely in contact with the surface of the can - this makes the normal force approach zero). Put $N = 0$ in equation **(1)**.

$\Rightarrow m \times g = (m \times v^2) / R.$

$\Rightarrow v = (R \times g)^{1/2}.$

But v is also equal to circumference divided by period of revolution $= (2 \times \pi \times R) / T$ where T is the period of revolution.

$\Rightarrow v = (R \times g)^{1/2} = (2 \times \pi \times R) / T.$

$\Rightarrow T = (2 \times \pi \times R) / (R \times g)^{1/2}.$

$\Rightarrow T = [2 \times \pi \times (0.4)] / (0.4 \times 10)^{1/2} = 1.26$ s.

Answer A!

29.   Centripetal acceleration is $a_c = v^2 / R$. Use $v = R \times \omega$. Then the formula for centripetal acceleration becomes: $a_c = (R \times \omega)^2 / R = R \times \omega^2$.

We need to know ω for the earth. Earth takes 24 hours to complete one revolution, so:

$\omega = \Delta\theta / t.$

$\omega = (2 \times \pi)$ radians $/ [24$ hours $\times (3600$ s $/ 1$ hour$)].$

Substitute back into the equation for centripetal acceleration and we get: $a_c = R \times \omega^2 =$
6400 km $\times (1000$ m $/ 1$ km$) \times \{(2 \times \pi)$ radians $/ [24$ hours $\times (3600$ s $/ 1$ hr$)]\}^2.$

$\Rightarrow a_c = [6400$ km $\times (1000$ m $/ 1$ km$)] \times \{2\pi / [(24 \times 3600)]\}^2.$

$\Rightarrow a_c = 3.4 \times 10^{-2}$ m/s$^2 = (3.4 \times 10^{-3} \times 10)$ m/s$^2.$

$\Rightarrow a_c = 3.4 \times 10^{-3} \times g.$

Answer A!

31.   The acceleration towards the center is $a_c = v^2 / R$. When an object moves in a circle at a constant speed, $v =$ circumference / period $= (2\pi \times R) / T.$

So $a_c = v^2 / R = [(2\pi \times R / T)]^2 / R = (4\pi^2 \times R) / T^2.$

With $R = 0.528 \times 10^{-10}$ m, and $T = 15 \times 10^{-15}$ s, we get $a_c = 9 \times 10^{18}$ m/s$^2.$

Answer A!

32.   The car travels from A to B. The distance traveled will be equal to the arc length traveled, which is equal to 314 meters. *Arc length* is found by multiplying the *radius* by the *angle* in radians $= R \times \theta$ where R is the radius in meters and θ is the angle in radians. Again, note that "Theta" is measured in radians.

Arc length $= 314$ m $= R \times 90$ degrees $\times \pi$ radians $/ 180$ degrees. So $R = (314 \times 180) / (90 \times \pi) = 200$ m.

The net force toward the center $= (m \times v^2) / R = F_c$

$= 1200$ kg $\times [(54$ km/hr $\times 1000$ m/km $\times 1$ hr/3600 s$)^2] / 200$ m. $\Rightarrow F_c = 1,350$ N.

Answer B!

34. When the earth goes around the sun, the centripetal force felt by the earth towards the center (of mass of the earth sun system) is $F_c = m \times v^2 / R$.

The centripetal force is provided by the gravitational attraction between the sun and the earth, $(G \times M_{earth} \times M_{sun}) / R^2_{e\,s}$.

$\Rightarrow (M_{earth} \times v^2) / R = (G \times M_{earth} \times M_{sun}) / R_{es}^2$.

We see that the mass of the earth is on both sides of the equation, so they cancel each other out leaving us with $v^2 R_{es} = G\, M_{sun} = $ a constant.

Let $R_{es} = R$. Now use $v = R \times \omega$:

$\Rightarrow [R^2 \times \omega^2 \times R] = $ constant. $\Rightarrow R^3 \times \omega^2$ is a constant.

We know that: $\omega = 2\pi / T$.

$\Rightarrow R^3 (2\pi / T)^2 = $ constant. $\Rightarrow R^3 / T^2 = $ constant as $4\pi^2$ is a constant.

We note here that a constant divided by another constant equals yet another constant.

$\Rightarrow T^2 \propto R^3$. This is an important result known as Kepler's Third Law of Planetary motion. This law was deduced observationally before Newton developed the laws of mechanics.

Answer B!

35. Gravitational attraction between the sun and the earth is what keeps the earth moving in circles (actually elliptic orbits) around the sun.

Answer A!

37. Earth makes one full rotation about its axis in 24 hours. Angular velocity is $\omega = \Delta\theta / t = $ (2π radians) / 24 hours. So $\omega = 2\pi$ radians / (24 hours × 1 hour / 3600 s).

$\Rightarrow \omega = 73 \times 10^{-6}$ rad/ s.

Answer A!

38. At the poles, and everywhere else, angular velocity is a constant $= 73 \times 10^{-6}$ rad/s.

Answer B!

40. At the top, net force is toward the center of the circle. The weight is toward the center of the circle and the normal force vertically up in the opposite direction. Remember that net force is toward the center.

Therefore,

$mg - N = (m \times v^2) / R$.

When the motorcycle barely makes it to the top, the normal force is zero (normal force can exist only if there is a surface to exert the force on the motorcycle). Set $N = 0$.

$\Rightarrow m \times g = (m \times v^2) / R. \Rightarrow g = v^2 / R. \Rightarrow v = (R \times g)^{1/2} = (160 \text{ m} \times 10 \text{ m/s}^2)^{1/2} = (1600 \text{ m}^2/\text{s}^2)^{1/2}$.

$\Rightarrow v = 40$ m/s. Now convert m/s to km/hr to get: $v = 40$ m/s × (3600 s / 1 hr) × (1 km / 1000 m) = 144 km/hr.

Answer B!

41. Gravitational attraction between two masses is $F_G = (G \times m_1 \times m_2) / R^2$.

$\Rightarrow F_G = (6.7 \times 10^{-11}) \times (9 \times 10^{-31}) \times (1.7 \times 10^{-27}) / (5 \times 10^{-11})^2$.

$\Rightarrow F_G = 4.1 \times 10^{-47}$ N.

Answer A!

43. $\omega = \Delta\theta / t = 2\pi$ radians $/ 1$ year $\times (1$ year $/ 365$ days$) = 0.017$ rad/day $\times (1$ day $/ 24$ hr$) \times (1$ hr $/ 3600$ s$)$.

$\omega = 2 \times 10^{-7}$ rad/s.

Note that the earth takes 365 days to go around the sun. One radian of a circularly shaped apple pie is a little less than 60 degrees. No wonder earth takes so much time.

Answer A!

44. The net acceleration towards the sun is v^2 / R. Use $v = R \times \omega$.

Therefore, $a_c = (R \times \omega)^2 / R = R \times \omega^2$. The distance from the earth to the sun $= 1.5 \times 10^{11}$ m (given in problem **23** of the book).

So $a_c = R \times \omega^2 = (1.5 \times 10^{11}) \times (2 \times 10^{-7})^2 = 6 \times 10^{-3}$ m/s^2.

Answer B!

46. As you move away from earth, the magnitude of gravity goes down. If you are outside the surface of the earth, $m \times g_p = (G \times m_E \times m) / r^2$.

$\Rightarrow g_p = (G \times m_E) / r^2$.

In this case, r is your distance from the earth's center. When you move away, as r increases, g decreases.

Answer A!

47. When you dive to the bottom of the ocean, you are closer to the center of the earth. Note that the water above you pulls you away from the center.

As the person goes down, gravity decreases because distance r is smaller.

This is correct because inside the earth, $g_{inside} = g_{surface} \times (r / R_{earth})$ where R_{earth} is the radius of the earth.

Answer A!

49. Poles are closer to the center of the earth than the equator is. Therefore, gravity at the poles will be higher than gravity at the equator. Gravity in the U.S. ≈ 9.8. Gravity at the poles $> 9.8 = 9.844$. Gravity at the equator $< 9.8 = 9.776$!

Alternate solution: The force between the earth and an object on the surface is $F = (Gm_{earth}m) / R_{earth}^2$. This force causes objects to fall, that is, to accelerate toward the earth. However, this force also provide the necessary centripetal force to keep the object traveling with the earth's rotation. Thus $F = (Gm_{earth}m) / R_{earth}^2 = (mg + ma_c)$. Since $(Gm_{earth}) / R_{earth}^2$ is a constant, g is smaller or greater as a_c is greater or smaller. At the pole, there is no translational velocity due to rotation, and a_c is zero. In fact, we observe at the pole $g_{Pole} = 9.844$ m/s^2. On the equator, a_c is maximum and g is smaller. Observe $g_c = 9.776$ m/s^2.

Answer B!

50. Weight will decrease when gravity is smaller. Gravity at the equator is smaller.
    Answer B!

52. We want the communication satellite to stay with us, so we make its period equal to one day. This is accomplished by choosing the orbital altitude appropriately.
    Answer A!

53. We expect g to decrease as $g \propto 1 / r^2$ outside the surface of the earth. Inside it follows the linear relation $g_{inside} = g_{surface} \times (r / R_{earth})$ where R_{earth} is the radius of the earth and r is the distance to any point within the surface of the earth..
    Answer D!

55. As the person goes down, gravity decreases because distance is smaller. This is correct because inside the earth, $g_{inside} = g_{surface} \times (r / R_{earth})$ where R_{earth} is the radius of the earth.
    Answer B!

56. $F = (G \times m_E \times m) / R^2$. This equation implies that gravitational force reduces as R increases. As R approaches ∞, this force approaches zero.
    Answer B!

58. Tension is the only force acting toward the center. $T = (m \times v^2) / R$.
    Answer D!

59. Tension and weight are both acting toward the center. $T + (m \times g) = (m \times v^2) / R$.
    Answer C!

61. $sin\ \alpha =$ opposite side to α / hypotenuse $= R / l$. $\Rightarrow$ Radius of circle $= l \times sin\ \alpha = l \times cos\ \theta = R$.
    Answer B!

62. Tension acts along the chord away from the object toward O.
    Along the horizontal: The net force towards the center is:
    $T\ cos\ \theta = T \times sin\ \alpha = m \times R \times \omega^2 =$
    $m \times (l \times sin\ \alpha) \times \omega^2$.
    $\Rightarrow T = m \times l \times \omega^2$.
    This is unnecessary, but illustrates that sometimes you start down a path that doesn't lead where you want to go.

    Along the vertical: Weight is balanced by the vertical component of the tension T.
    $\Rightarrow T\ sin\ \theta = T \times cos\ \alpha = mg$. $\Rightarrow T = mg / cos\ \alpha$.
    Answer C!

64.   Tension will go up.

From problem **62**, $T = mg / \cos \alpha$. As $\cos \alpha$ decreases in value, T goes up.

Answer B!

65.   Weight is balanced by the vertical component of the tension T.

$\Rightarrow T \sin \theta = T \times \cos \alpha = m \times g.$ **(1)**

The horizontal component of the tension provides the centripetal force.

$\Rightarrow T \cos \theta = T \times \sin \alpha = m \times R \times \omega^2 = m \times (l \times \sin \alpha) \times \omega^2.$ $\Rightarrow T = m \times l \times \omega^2.$ **(2)**

Substitute for T in equation **(1)**. $\Rightarrow [m \times l \times \omega^2] \times \cos \alpha = m \times g.$

$\Rightarrow \omega^2 = g / (l \times \cos \alpha) = g / h.$

Period $= 2\pi / \omega$. We find $\omega = [g / (l \cos \alpha)]^{1/2}$.

So period, $T = 2\pi / [g / (l \times \cos \alpha)]^{1/2}$.

$\Rightarrow$ Period of revolution $= 2\pi [(l \times \cos \alpha) / g]^{1/2}$.

Answer C!

67.   The gravity on the surface of the earth is $g_E = G \times (M_{earth} / r_{Earth}^2) \approx 10$ m/s^2.

The centripetal acceleration is

$a_c = v_{Earth}^2 / r_{Earth} = r_{Eareth} \times \omega_{Earth}^2$.

The angular speed of the earth $= \Delta \theta / t = 2\pi / [1 \text{ day} \times 24 \text{ hr} / \text{day} \times 3{,}600 \text{ s} / 1 \text{ hr}] = 7.3 \times 10^{-5}$ rad/s. $\Rightarrow$ $a_C = r_E \times \omega^2$.

$\Rightarrow a_c = (6.37 \times 10^6) \times (7.3 \times 10^{-5})^2 = 0.034$ m/s^2.

The ratio of centripetal acceleration to the gravitational acceleration $= 0.034 / 10 = 0.0034 = 3.4 \times 10^{-3}$.

Answer B!

68.   It should be exactly ONE!

The gravity of earth provides the centripetal force to keep the moon in orbit!

The moon takes 28 days to go around the earth.

The angular speed of the moon is $\omega_{Moon} = \Delta \theta / t =$

$2\pi / [(28 \text{ days} \times 24 \text{ hr per day} \times 60 \text{ min per hr} \times 60 \text{ s per min}) = 2.6 \times 10^{-6}$.

The centripetal acceleration felt by the moon $= r \times \omega^2 = (3.8 \times 10^8) \times (2.6 \times 10^{-6})^2 = 25.7 \times 10^{-4}$.

The centripetal force is $m_{Moon} \times 25.7 \times 10^{-4}$ newtons.

The gravitational force between the earth and moon $=$

$(G \times m_{Earth} \times m_{Moon}) / (3.8 \times 10^8)^2 = (6.7 \times 10^{-11} \times 6 \times 10^{24} \times m_{Moon}) \times (0.07 \times 10^{-16}) = 2.8 \times 10^{-3} \times$

$m_{Moon} = 28 \times 10^{-4} \times m_{Moon}$. The ratio is 1 (ONE).

Answer A!

70. $g_E = (G \times M_E) / r_E^2$ and $g_h = (G \times M_E) / (r_E + h)^2$. $\Rightarrow g_E / g_h = (r_E + h)^2 / r_E^2$. We expect gravitational pull to increase on the way down. Note that the subscript E stands for earth.

Answer B!

71. For the earth satellite, radius of the orbit $\approx R_e$. If m is its mass of the satellite, the centripetal force is provided by the gravitational force.

$\Rightarrow (G \times m \times M_e) / R_e^2 = (m \times v^2) / R_e.$

But $(G \times M_e / R_e^2) = g_E.$

$\Rightarrow \ g_E = v^2 / R_e.$

$\Rightarrow \ v^2 = g_E \times R_e \ \text{or} \ v = (g_E \times R_e)^{1/2}.$

Answer B!

1.   Answer A!   2.   Answer C!

4.   Answer B!   5.   Answer A!

7.   Answer A!   8.   Answer B!

10.   Answer B!   11.   Answer A!

13.   Answer C!   14.   Answer A!

16.   Answer B!   17.   Answer B!

19.   Answer B!   20.   Answer C!

22.   Answer B!   23.   Answer B!

25.   Answer B!   26.   Answer B!

28.   Answer B!   29.   Answer A!

31.   There is no displacement. She is waiting.
        Answer A!

32.   The velocity of the suitcase does not change. This means that the work done is zero.
        Answer C!

34.   The suitcase is accelerating. There has to be net work done on the system.
        Net Work = $F_{\text{NET FORCE}} \times s_{\text{DISPLACEMENT}} \times cos\,\theta$ ANGLE BETWEEN THE DIRECTION OF FORCE AND DIRECTION OF

        DISPLACEMENT.
        STEP 1:   Net work done = $m_{\text{MASS}} \times a_{\text{ACCELERATION}} \times s_{\text{DISPLACEMENT}} \times cos\,0$, where m = 15 kg and s is
                         the displacement which = 50 m. Note that the angle between the direction of net force and
                         direction of displacement is $0°$ because the system is accelerating, that is, the direction of
                         acceleration and displacement vectors are in the same direction.
        STEP 2:   Find acceleration using equations of kinematics. We need to find acceleration. Use
                         $\Delta x_{\text{DISPLACEMENT}} = [v_{0x,\text{INITIAL VELOCITY}} \times t_{\text{TIME}}] + [\frac{1}{2}\,a_{x,\text{ ACCELERATION}} \times t_{\text{TIME}}^{2}]$.
        Assuming the person starts with a zero velocity, we get:
        $\Delta x = \frac{1}{2} \times a_x \times t^2$.
        $\Rightarrow 50_{\text{DISPLACEMENT}} = \frac{1}{2}\,a\,(15_{\text{TIME}})^2$.
        $\Rightarrow a_{\text{ACCELERATION}} = [2\,(50)] / 15^2 = 0.44$ m/s^2.

Net work done $W = m_{MASS} \times a_{ACCELERATION} \times s_{DISPLACEMENT} \times cos\ 0 =$
$15_{MASS} \times 0.44_{ACCELERATION} \times 50_{DISPLACEMENT} = 333$ Joules.
Answer A!

35. At point A:

1. $KE = \frac{1}{2}\ m_{MASS} \times v_{0,\ INITIAL\ VELOCITY}^2 =$ Kinetic Energy (T).

2. $PE = m_{MASS} \times g_{GRAV.} \times y_{A,\ VERTICAL\ HEIGHT\ OF\ A\ FROM\ THE\ ZERO\ LEVEL} = 0$ as height of A from point A
is 0 m if A is chosen as the zero level for the potential energy.

Total Energy, E = Kinetic Energy (T) + Potential Energy (V).
$E_{TOTAL\ ENERGY} = T_{KINETIC\ ENERGY} + V_{POTENTIAL\ ENERGY}.$
$\Rightarrow E = T$ because $V = 0$ at point A $= \frac{1}{2}\ m_{MASS} \times v_{0,\ INITIAL\ VELOCITY}^2.$
So, kinetic energy at A $= \frac{1}{2}\ m \times v_0^2$; potential energy at A $= 0$; and so total energy at A is all kinetic as
potential energy at A is zero.

3. $E_{TOTAL\ ENERGY} = \frac{1}{2}\ m_{MASS} \times v_{0,\ INITIAL\ VELOCITY}^2.$
Answer B!

37. Point B is the maximum height.
KE at B $= \frac{1}{2}\ m_{MASS} \times v_{B,\ VELOCITY}^2$ where v_B is the velocity at point B.
Velocity at point B is strictly horizontal velocity, $v_0\ cos\ \alpha$ as point B is at the maximum height where vertical
velocity along y is zero. This is a very important concept.
The square of a velocity, $v_{VELOCITY}^2 = v_{HORIZONTAL\ COMPONENT}^2 + v_{VERTICAL\ COMPONENT}^2.$
The vertical component of the velocity at the maximum height is zero. That is, the kinetic energy at the
maximum height will be zero only if the horizontal component of the velocity is also zero! Do not forget the
last sentence. Read it again and proceed only if you understand that step! In general,
$v_{VELOCITY}^2 = v_{HORIZONTAL\ COMPONENT}^2 + v_{VERTICAL\ COMPONENT}^2.$

SOLUTION TO OUR PROBLEM:
$v_{HORIZONTAL\ COMPONENT} = v_0\ cos\ \alpha.$
$v_{VERTICAL\ COMPONENT} = 0$ at the maximum height.
$v_{VELOCITY}^2 = (v_0\ cos\ \alpha)^2 + 0 = v_0^2\ cos^2\ \alpha.$
Note that this particular solution is true only for the maximum height.
This solution makes kinetic energy at point B $= \frac{1}{2}\ m \times v_0^2\ cos^2\ \alpha.$

$\Rightarrow KE$ at point B $= \frac{1}{2}\ m \times (v_0\ cos\ \alpha)^2 = \frac{1}{2}\ m \times v_0^2\ cos^2\ \alpha.$　　　　　**(1)**.
PE at point B $= m_{MASS} \times g_{GRAV.} \times h_{B,\ VERTICAL\ HEIGHT\ OF\ B\ FROM\ ZERO\ LEVEL}$, where h_B is the height of point B
from the zero level.
$\Rightarrow PE$ at point B $= m \times g \times h_{max}.$　　　　　**(2)**.
TE at point B $= \frac{1}{2}\ m \times v_0^2$ (see problem **35**).　　　　　**(3)**.
Answer C!

38.   At point C:

$v_{\text{VELOCITY}}^2 = v_{\text{HORIZONTAL COMPONENT}}^2 + v_{\text{VERTICAL COMPONENT}}^2$.

SOLUTION TO OUR PROBLEM AT POINT C:

$v_{\text{HORIZONTAL COMPONENT}} = v_0 \cos \alpha$.

Acceleration component along the horizontal is zero.

$v_{\text{VERTICAL COMPONENT}} = -(v_0 \sin \alpha)$. Negative because the projectile is on the way down. It is $(v_0 \sin \alpha)$ because vertical velocities at the same level have equal magnitudes but opposite directions. Point C is at the same horizontal level as point A, the starting point. We are sure you know by now that the vertical component of the velocity at the start is $+(v_0 \sin \alpha)$. Positive because it is on the way up. This makes the vertical velocity at point C which is at the same height as point A = $-(v_0 \sin \alpha)$.

$v_{\text{VERTICAL COMPONENT}}^2 = (-v_0 \sin \alpha)^2 = v_0^2 \sin^2 \alpha$.

KE at C = $\frac{1}{2} m \times v_C^2$ where v_C = velocity at point C.

$v_C^2 = v_{Cx}^2 + v_{Cy}^2$ where v_{Cx} = velocity at C along x and v_{Cy} = velocity at point C along y.

$v_C^2 = (v_0 \times \cos \alpha)^2 + (-v_0 \times \sin \alpha)^2$ or $v_C^2 = v_0^2$.

$\Rightarrow KE$ KINETIC ENERGY at C = $[\frac{1}{2} m \times v_0^2]$.

PE POTENTIAL ENERGY at C = 0 as vertical displacement of point C is zero from the zero level.

TE TOTAL ENERGY = KE KINETIC ENERGY + PE POTENTIAL ENERGY =

       $[\frac{1}{2} m \times v_0^2]$ KINETIC ENERGY + $[0]$ POTENTIAL ENERGY = $\frac{1}{2} m \times v_0^2$.

Answer D!

40. KE decreases with increasing height, but is not zero at maximum height, $h_{\max}$ (if α is not 90 degrees). Details: At time $t = 0$, $KE_{\text{KINETIC ENERGY}}$ is not zero (for the projectile). The question is will kinetic energy ever become zero for this particular problem. Answer is never! At the maximum height is where we should expect maximum potential energy but minimum kinetic energy. In problem **37**, we proved that the kinetic energy is not zero at the maximum height. Do not forget that at the maximum height, vertical velocity is zero but horizontal velocity stays the same. Beyond the maximum height, the contribution of the kinetic energy part of the total energy increases. Because of the square term, we do not expect this to be a straight line if horizontal displacement x is plotted with kinetic energy.

Answer A!

41.   Total Energy is a constant if there is no air resistance. The contribution to the total energy (kinetic and potential energies) will vary due to position but not the total energy.

Answer A!

43.   When $x = R / 2$, vertical displacement (height) is maximum as y is maximum. Note that R is the range.

Answer B!

44. $KE_{\text{KINETIC ENERGY}}$ is maximum when $PE_{\text{POTENTIAL ENERGY}}$ is minimum. In our case, PE is minimum at the starting and the ending point. For this particular problem, points A and C will have maximum kinetic energy but minimum potential energy. Again, remember that points A and C are at the same (zero) level.
Answer A!

46. At point A, Total Energy TE or E is purely *kinetic*. If TE is all PE, then $KE = 0$. At maximum height, velocity along y is zero *but* velocity along x is not. For this reason, Kinetic Energy (T) or KE can never be zero. That is, for our problem, for a projectile kicked at an angle to the horizontal, total energy can never be all potential energy as the non–zero horizontal velocity makes the kinetic energy a non–zero number. Again, we are reminding you that the horizontal velocity never goes to zero. It stays the same for a projectile as long as there is no acceleration along the horizontal.
Answer B!

47. Potential Energy, $V = m_{\text{MASS}} \times g_{\text{GRAV.}} \times h_{\text{VERTICAL HEIGHT FROM THE ZERO LEVEL}} = m \times g \times y$. We expect this to increase until we reach maximum height and then decrease back to zero.
$V = PE_{\text{POTENTIAL ENERGY}} = m \times g \times y$.
$\Rightarrow V / y = m \times g = \text{slope} = \text{constant}$.
$\Rightarrow V_{\text{POTENTIAL ENERGY}}$ versus vertical displacement y is a straight line graph with the magnitude of the slope $= m_{\text{MASS}} g_{\text{GRAV.}}$.
Answer B!

49. $TE_{\text{TOTAL ENERGY}} = KE_{\text{KINETIC ENERGY}} + PE_{\text{POTENTIAL ENERGY}}$.
$\Rightarrow E_{\text{TOTAL ENERGY}} = T_{\text{KINETIC ENERGY}} + V_{\text{POTENTIAL ENERGY}}$.
At the starting point, potential energy $V = m_{\text{MASS}} \times g_{\text{GRAV.}} \times h_{\text{VERTICAL DISPLACEMENT FROM ZERO LEVEL}} = m \times g \times y = 0$. That is, at the start, total energy is all kinetic. $\Rightarrow E = T$. If there is no friction, this total energy will stay the same (even though kinetic energy and potential energy may vary) as energy is conserved.
Answer A!

50. Read the solution to problem **37**. Vertical component of the velocity at the maximum height is zero. Horizontal component of the velocity at maximum height is ($v_0 \times \cos \alpha$).
$v_{\text{HORIZONTAL COMPONENT}} = (v_0 \cos \alpha)$.
$v_{\text{VERTICAL COMPONENT}} = 0$ at the maximum height.
$v_{\text{VELOCITY}}^2 = (v_0 \cos \alpha)^2 + 0 = v_0^2 \cos^2 \alpha$. Note that this particular solution is true only for the maximum height. This solution makes kinetic energy at point B $= \frac{1}{2} m \times v_0^2 \cos^2 \alpha$.
KE at the maximum height $= \frac{1}{2} m \times (v_0 \times \cos \alpha)^2$. Remember that velocity along y at maximum height $= 0$.
Point B represents $\frac{1}{2} m \times v_0^2 \times \cos^2 \alpha$. This is represented by BM on the graph. This is actually the minimum value of kinetic energy (because potential energy is maximum).

At the start at point A, Total Energy (which is the sum of kinetic and potential energies) is all kinetic because potential energy is zero at point A.
$TE_{\text{TOTAL ENERGY}}$ at A $= KE_{\text{KINETIC ENERGY}}$ at A $= [\frac{1}{2} m \times v_0^2]_{\text{KINETIC ENERGY}}$. This is represented by AO.

AO represents the Total Energy at the start.

$\Rightarrow \text{BM} / \text{AO} = (\frac{1}{2}\, m \times v_0^2 \times cos^2\, \alpha) / (\frac{1}{2}\, m \times v_0^2) = cos^2\, \alpha.$

Answer B!

52.   The maximum value of $PE_{\text{POTENTIAL ENERGY}}$ occurs at the maximum height from the zero level.

$TE_{\text{TOTAL ENERGY}}$ at max height $= [KE_{\text{KINETIC ENERGY}} + PE_{\text{POTENTIAL ENERGY}}]$ at max height.

$\Rightarrow [\frac{1}{2}\, m \times v_0^2]_{\text{TOTAL ENERGY}} = [\frac{1}{2}\, m \times v_0^2 \times cos^2\, \alpha]_{\text{KINETIC ENERGY}} + PE_{\text{max}}.$

$\Rightarrow PE_{\text{max}} = [\frac{1}{2}\, m \times v_0^2]_{\text{TOTAL ENERGY}} - [\frac{1}{2}\, m \times v_0^2 \times cos^2\, \alpha]_{\text{KINETIC ENERGY AT MAX HEIGHT}}.$

$\Rightarrow PE_{\text{max}} = [(\frac{1}{2}\, m \times v_0^2) \times (1 - cos^2\, \alpha)]$

Use the trig relation: $(1 - cos^2\, \alpha) = sin^2\, \alpha.$

$\Rightarrow \quad PE_{\text{max}} = \frac{1}{2}\, m \times v_0^2 \times sin^2\, \alpha.$

Note that $1 = sin^2\, \alpha + cos^2\, \alpha. \quad \Rightarrow \quad sin^2\, \alpha = 1 - cos^2\, \alpha.$

So $\frac{1}{2}\, m \times v_0^2 (1 - cos^2\, \alpha) = \frac{1}{2}\, m \times v_0^2 \times sin^2\, \alpha.$

Answer B!

53.   If you choose zero level at the starting point (that is what we have been doing so far), then the answer is A! We expect the *PE* to increase, become a maximum at the maximum height, and then decrease back to zero. Answer A!

55.   By energy conservation, total energy (the sum of kinetic and potential energies) remains a constant if there is no air resistance.

Answer A!

56.   OA represents the value of Kinetic Energy at the start at $t = 0$ seconds, which equals $\frac{1}{2}\, m \times v_0^2.$

Answer A!

58.   Answer B!

59.   It is the Total Energy which is the sum of kinetic and potential energies that remains a constant.

Answer C!

61.   Let BC be the height *h*.

Let AC be the distance *s*.

The work done against gravity to take an object to point C is the same for both paths AC and BC.

Note that gravitational force is a conservative force and so work done is independent of path. The work is the same. You are doing work against gravity. However, the forces are different because the distances are not the same. (I do not have to tell you that).

When we use an inclined plane to slide a carpet, we use a smaller force but a bigger distance. When we lift the carpet vertically, the force is bigger but distance is smaller. If work done is the same against gravity for both the paths then:

$F_{\text{BIGGER FORCE}} \times d_{\text{SMALLER DISTANCE}} = f_{\text{SMALLER FORCE}} \times D_{\text{BIGGER DISTANCE}}$.

$[m \times g]_{\text{BIGGER FORCE}} \times [h_{\text{SMALLER VERTICAL DISTANCE}}] = [m \times g \times \sin\theta]_{\text{SMALLER FORCE}} \times s_{\text{BIGGER DISTANCE AC}}$

Note that vertical height $h = s_{\text{DISTANCE AC}} \times \sin\theta_{\text{ANGLE OF THE INCLINED PLANE WITH THE HORIZONTAL}}$.
[By definition, $\sin\theta$ = opposite side to θ / hypotenuse. $\Rightarrow \sin\theta = h / s$.]
Answer A!

62. $Area = F_{\text{PLOTTED ON } y \text{ AXIS}} \times displacement_{\text{PLOTTED ON } x \text{ AXIS}}$.
This is nothing but the area under the curve which is actually the work done.
Answer B!

64. When a body moves down an inclined plane, work is done by gravity – actually by the component of gravity along the inclined plane. The normal force acts perpendicular to the direction of displacement, so no work will be done by the normal force. We can eliminate part C because when we slide down, work is done by the field rather than against gravity. When we climb up, we do work against gravity.
Answer A!

65. Work done by center seeking forces is zero, as the net force is always acting towards the center while the displacement is always perpendicular to it. The net force is always acting perpendicular to the direction of displacement. This makes the angle between the direction of force and the direction of displacement 90 degrees. $W = F \times s \times \cos\theta = F \times s \times \cos 90 = 0$ as $\cos 90 = 0$. That is, the centripetal force does not contribute to the work.
Answer C!

67. $Power = work / time \Rightarrow Power = Force$ on the horizontal road $\times velocity$. Velocity is a constant. That means, force is nothing but the force to overcome friction to keep the velocity a constant.
$Power = P, Force = \mu \times m \times g$.
$\Rightarrow P = \mu \times m \times g \times v$.
$\Rightarrow v = P / (\mu \times m \times g)$.
Answer B!

68. When you climb up, you are doing work against gravity and friction. Again, you are told that the truck is climbing up at constant velocity. That means, acceleration is zero. Or sum of the forces acting down towards the base of the inclined plane is equal to the sum of the forces provided by the truck to climb up at constant velocity. Both frictional force and the gravitational force component are towards the base of the inclined plane.
Net force down = $\mu \times m \times g \times \cos\alpha + m \times g \times \sin\alpha$. If truck has to go up the inclined plane at constant velocity, it must provide this force in the opposite direction, up the inclined plane. From a unit point of view:
$P = Power = Work / Time = (F \times distance) / time = F \times (distance / time)$.
$\Rightarrow Power = F_{\text{NET FORCE APPLIED BY THE TRUCK}} \times v$.

$\Rightarrow Power = (\mu \times m \times g \times \cos \alpha + m \times g \times \sin \alpha) \, v.$

$\Rightarrow v = P / (\mu \times m \times g \times \cos \alpha + m \times g \times \sin \alpha) = P / [m \times g \, (\mu \times \cos \alpha + \sin \alpha)].$

Answer A!

70. $[\frac{1}{2} \, m \times v_{\text{VELOCITY}}^2]_{\text{KINETIC ENERGY}}$ is always positive as long as we never invent or discover a particle with a negative mass. Note that velocity can be negative but its square is always positive.

Answer B!

71. $[KE_{\text{KINETIC ENERGY}} + PE_{\text{POTENTIAL ENERGY}}]$ = TOTAL ENERGY. If kinetic energy is zero, total energy takes the form of potential energy. By definition, $PE = V = m_{\text{MASS}} \times g_{\text{GRAV.}} \times h_{\text{VERTICAL DISPLACEMENT FROM THE ZERO LEVEL}}$. The vertical displacement can be positive (above the zero level), zero (at the zero level) or even negative (below zero level). That is, total energy can be positive, zero, or negative.

Answer A!

73. Note that momentum, p, is equal to $m \times v$. Kinetic energy, $KE = \frac{1}{2} \, m_{\text{MASS}} \times v_{\text{VELOCITY}}^2 = \frac{1}{2} \, m \times (p / m)^2$.

$KE_{\text{KINETIC ENERGY}} = p_{\text{MOMENTUM}}^2 / (2 \, m_{\text{MASS}})$.

Answer B!

74. Kinetic energy, $KE = \frac{1}{2} \, m_{\text{MASS}} \times v_{\text{VELOCITY}}^2$. When $v = 0$, $KE = 0$. When v is non-zero, KE is positive. Because of the v^2 term involving velocity, KE can never be negative.

Answer A!

76. $TE = KE + PE$. If $KE = 0$, Total Energy takes the form of PE.

By definition, $PE = V = m_{\text{MASS}} \times g_{\text{GRAV.}} \times h_{\text{VERTICAL DISPLACEMENT FROM THE ZERO LEVEL}}$. The vertical displacement can be positive (above the zero level), zero (at the zero level) or even negative (below zero level). Because PE can be zero, positive or negative, TE can be positive, zero, or negative.

Answer C!

77. For non-zero speed, we expect KE to be greater than zero. $TE = KE + PE$ or $E = T + V$.

$KE = TE - PE$ or $T_{\text{KINETIC ENERGY}} = E_{\text{TOTAL ENERGY}} - V_{\text{POTENTIAL ENERGY}}$.

For T to be a non-zero number we need to have a non-zero speed.

TE has to be greater than Potential Energy.

Answer A!

79. $PE = m \times g \times h = m \times g \times BD = m \times g \times (AB - AD) = m \times g \times (l - l \times \cos \theta) = m \times g \times l \times (1 - \cos \theta)$.

Note that in the right angled triangle (one of the angles is $90°$) ADC: $\cos \theta = AD / AC$.

Or $\cos \theta = AD / l$. Note that AC is the radius $l = AB$.

Answer A!

80. In the previous problem, we took the length l as the length to the center of mass of the body. Here in this problem, the distance to the center of mass, $l = L / 2$.

The work is done against gravity.

$\Rightarrow PE = m \times g \times h = [m \times g] \times [(L / 2) \times (1 - cos\,\theta)]_{\text{VERTICAL DISPLACEMENT } h \text{ FROM THE ZERO LEVEL}}$.

Answer B!

82. Work done by gravity from A to B is zero, since the force of gravity is acting down and displacement is to the right. The angle between the direction of the force of gravity and the direction of displacement is 90 degrees.

From A to B, $W_{AB} = F \times s \times cos\,90 = 0$.

From B to C, $W_{BC} = (m \times g) \times (2\,m) \times cos\,180 = -m \times g \times (2) =$

$-40_{\text{ MASS IN kg}} \times 10_{\text{GRAV.}} \times 2_{\text{ VERTICAL DISPLACEMENT}} = -800$ Joules.

Total work done = -800 Joules. (The negative is because we calculated the work done by gravity. Lifting the box, we did work against gravity. Its magnitude = 800 joules.

Answer B!

83. Now we have to do work against friction and work against gravity. Let us just calculate the magnitude.

W_{net} = Net work done = $[(F_{f,\text{ FORCE OF FRICTION}}) \times d_{\text{DISTANCE}}]_{A \text{ to } B} + [(m \times g) \times h]_{B \text{ to } C} =$

$[(\mu \times m \times g)_{\text{FORCE OF FRICTION}} \times d]_{\text{WORK DONE AGAINST FRICTION}} + (m \times g \times h)_{\text{WORK DONE AGAINST GRAVITY}}$.

Net Work Done =

$(0.3_{\text{ COEFFICIENT OF FRICTION}} \times 40_{\text{MASS}} \times 10_{\text{GRAV.}} \times 5_{\text{ DISTANCE ALONG WHICH THE FORCE ACTS}}) +$

$(40_{\text{ MASS}} \times 10_{\text{ GRAV.}} \times 2_{\text{ VERTICAL DISPLACEMENT}}) = 600 + 800 = 1400$ J.

Answer A!

85. *Work* = $m \times g \times h$, where h is the distance through which the center of mass has to be raised against gravity.

$= m \times g \times l / 2$.

Answer B!

86. When Total Energy is all Potential Energy at the maximum height, $KE = 0$ if thrown vertically up!

$KE_{\text{ KINETIC ENERGY AT THE START}} + PE_{\text{ POTENTIAL ENERGY AT THE START}} =$

$KE_{\text{ KINETIC ENERGY AT THE END}} + PE_{\text{ POTENTIAL ENERGY AT THE END}}$.

$[\frac{1}{2}\,m_{\text{MASS}} \times v_{\text{VELOCITY}}^2]_{\text{KINETIC ENERGY AT THE START}} + 0_{\text{POTENTIAL ENERGY AT THE START}} =$

$0_{\text{KINETIC ENERGY AT THE END}} + [m_{\text{MASS}}\,g_{\text{GRAV.}}\,h_{\text{VERTICAL HEIGHT}}]_{\text{POTENTIAL ENERGY AT THE END}}$.

Answer C!

88. Gravitational force is a conservative force. If you go from A to B, you gain *PE*. When you go back to A, you lose the energy you gained.

If on the other hand, you go from A to B, you lose energy if friction is present. On the way back you again lose energy due to friction.

Answer A!

89.　Work done by gravitational force (a conservative force) is independent of the path.
Answer D!

91.　Set total energy at the instant of rebounding = total energy after rebounding at the maximum height.

$KE_{\text{KINETIC ENERGY AT THE START}} + PE_{\text{POTENTIAL ENERGY AT THE START}} =$

$KE_{\text{KINETIC ENERGY AT THE END}} + PE_{\text{POTENTIAL ENERGY AT THE END}}.$

$[\frac{1}{2}\, m_{\text{MASS}} \times v_{\text{VELOCITY}}^2]_{\text{KINETIC ENERGY AT THE START}} + 0_{\text{POTENTIAL ENERGY AT THE START}} =$

$0_{\text{KINETIC ENERGY AT THE END}} + [m_{\text{MASS}}\, g_{\text{GRAV.}}\, h_{\text{VERTICAL HEIGHT}}]_{\text{POTENTIAL ENERGY AT THE END}}.$

TE after rebounding $= \frac{1}{2}\, m \times v^2.$

TE at maximum height after rebounding $= m \times g \times (3\, h\, /\, 4).$

Solve for $v \Rightarrow v =$ square root of $[2\, g \times (3\, h\, /\, 4)] =$ square root of $[(3\, g \times h\,)\, /\, 2].$

Answer C!

1. For the system in consideration, there should be no external force. For example, if you have two objects colliding, there *is* a net force on each of the individual colliding bodies at the instant they collide. (We have a change in speed, hence a change in velocity, hence an acceleration, hence a force...) But, if you look at the *two* masses as *a* system, the force on one due to the other is equal and opposite and acts only within the system among its parts.
Answer B!

2. As we said before in problem **1**, it is for the *system* (in this case the system consisting of two colliding masses) that the momentum is conserved and *not* for the individual bodies colliding. The law of conservation of momentum applies for *both* masses, which is actually the system.
Answer A!

4. If you define the system with *both* masses, both energy and vector sum of momentum have to be conserved (They have to be same before and after collision!) By setting vector sum of momentum before and after collision, we get:

$[(m_1$,MASS OF FIRST OBJECT v_{1i}, INITIAL VELOCITY OF FIRST OBJECT BEFORE COLLISION$) +$
$(m_2$,MASS OF SECOND OBJECT v_{2i}, INITIAL VELOCITY OF SECOND OBJECT BEFORE COLLISION $)]$VECTOR SUM OF MOMENTUM BEFORE COLLISION $=$
$[(m_1$,MASS OF FIRST OBJECT v_{1f}, FINAL VELOCITY OF FIRST OBJECT AFTER COLLISION $) +$
$(m_2$,MASS OF SECOND OBJECT v_{2f}, FINAL VELOCITY OF SECOND OBJECT AFTER COLLISION$)]$VECTOR SUM OF MOMENTUM AFTER COLLISION.

If you don't like all the subscripts, here it is, again! $m_1v_{1i} + m_2v_{2i} = m_1v_{1f} + m_2v_{2f}$.
By the energy conservation equation, we get:
$[(½)m_1v_{1i}{}^2$]KINETIC ENERGY OF FIRST OBJECT BEFORE COLLISION $+$
         $[(½)m_2v_{2i}{}^2$]KINETIC ENERGY OF SECOND OBJECT BEFORE COLLISION $=$
$[(½)m_1v_{1f}{}^2$]KINETIC ENERGY OF FIRST OBJECT AFTER COLLISION $+$
         $[(½)m_2v_{2f}{}^2$]KINETIC ENERGY OF SECOND OBJECT AFTER COLLISION
Answer B!

5. x_{CM}, POSITION OF CENTER OF MASS $=$
$(m_1$, MASS $\times x_1$, COORDINATE $+ m_2$, MASS $\times x_2$, COORDINATE$) / (m_1$, MASS $+ m_2$, MASS$)$.
Note that x_1 and x_2 (the position of masses) are coordinates.
These can be positive (to the right of origin), zero (at the origin) or negative (to the left of origin).
Answer A!

7. Velocity of center of mass $= (m_1$, MASS $\times v_1$, VELOCITY $+ m_2$, MASS $\times v_2$, VELOCITY$) / (m_1$, MASS $+ m_2$, MASS$)$.
Answer A!

8.   STEP 1:   The important thing to understand is that both velocities have the same sign because they are moving in the same direction.

      STEP 2:   Velocity of the center of mass =
                 $(m_{1, \text{MASS}} \times v_{1, \text{VELOCITY}} + m_{2, \text{MASS}} \times v_{2, \text{VELOCITY}}) / (m_{1, \text{MASS}} + m_{2, \text{MASS}})$.

                 Velocity of the center of mass = $(m_1 \times v_1 + m_2 \times v_2) / (m_1 + m_2)$.
                 $\{[5_{\text{MASS}} \times (+ 8)_{\text{VELOCITY}}] + [10_{\text{MASS}} \times (+2)_{\text{VELOCITY}}]\} / (5_{\text{MASS}} + 10_{\text{MASS}})$
                 = 4 m/s = +4 m/s (in the same direction as the 5 gram body because of the sign!). No change in the sign.

      Answer A!

10.   It is due to the principle of conservation of linear momentum that there is no change in the velocity of the center of mass during collision.
      In all collisions, whether elastic or inelastic, the linear momentum is conserved. As we said before, if you take the two masses as the system, net force has to be zero as force on mass m_1 due to m_2 is equal and opposite to the force on mass m_2 by m_1. Therefore net force is zero on the system (if you consider the two masses as the system). This can be true only if velocity of the center of mass is a constant before and after collision!
      Note that if your system has only mass m_1, net force is not zero.
      Similarly if you take m_2 as the system, net force is not zero. Just reminding you again, that's all!
      Answer D!

11.   Conservation of momentum applies to all collisions. That is the equation we learned before for conservation of linear momentum and it is true for both elastic and inelastic collisions. If you have forgotten it, here it is!
      $[(m_{1,\text{MASS OF FIRST OBJECT}} \, v_{1i, \text{INITIAL VELOCITY OF FIRST OBJECT BEFORE COLLISION}}) +$
      $\quad (m_{2,\text{MASS OF SECOND OBJECT}} \, v_{2i, \text{INITIAL VELOCITY OF SECOND OBJECT BEFORE COLLISION}})]_{\text{VECTOR SUM OF}}$
      $\quad \text{MOMENTUM BEFORE COLLISION}} =$
      $[(m_{1,\text{MASS OF FIRST OBJECT}} \, v_{1f, \text{FINAL VELOCITY OF FIRST OBJECT AFTER COLLISION}}) +$
      $\quad (m_{2,\text{MASS OF SECOND OBJECT}} \, v_{2f, \text{FINAL VELOCITY OF SECOND OBJECT AFTER COLLISION}})]_{\text{VECTOR SUM OF}}$
      $\quad \text{MOMENTUM AFTER COLLISION}}.$

      If you don't like all the subscripts, here it is, again! $m_1 v_{1i} + m_2 v_{2i} = m_1 v_{1f} + m_2 v_{2f}$.
      Note that energy is conserved only for elastic collisions.
      Answer B!

13.   Since $m_{\text{EJECTED MASS PER s}}$ and $v_{\text{VELOCITY OF THE EJECTED MASS}}$ remain constant until burn out, the thrust remains a constant at $m_{\text{EJECTED MASS PER s}} \times v_{\text{VELOCITY OF THE EJECTED MASS}}$.

      Note that the unit of thrust is: $[(\text{kg/s}) \times (\text{m/s}) = (\text{kg} \times \text{m/s}^2) = \text{newtons}!$ Hmmm! This is the unit of force!

Also, the gases ejected go one way and the rocket the opposite way. The ejected gases have a velocity. The ejected gases have a mass.

Hence, the ejected gases carry momentum (p MOMENTUM $= m$ MASS v VELOCITY) with them.

Answer C!

14.   At any instant if M is the mass of the rocket, then for the rocket, we have:

[T THRUST]BIGGER FORCE $-$ [M MASS OF ROCKET $\times$ g GRAV.]SMALLER FORCE $=$
       [M MASS OF ROCKET $\times$ a ACCELERATION]NET FORCE.

Hence, a ACCELERATION $=$ [T THRUST$/M$ MASS OF ROCKET] $-$ [g GRAV.].

Now substitute for thrust.

a ACCELERATION $=$
[(m EJECTED MASS PER s $\times$ v VELOCITY OF THE EJECTED MASS) $/M$MASS OF ROCKET] $-$ [gGRAV.].

Since MMASS OF ROCKET is decreasing when the fuel is ejected, the acceleration a increases until the fuel is completely burned out.

Answer A!

16.   The momentum of the gun is equal and opposite to momentum of the gun. Note that the momentum before we fire the gun is zero.

[($m_{1,}$MASS OF FIRST OBJECT $v_{1i,}$ INITIAL VELOCITY OF FIRST OBJECT BEFORE COLLISION) $+$
($m_{2,}$MASS OF SECOND OBJECT $v_{2i,}$ INITIAL VELOCITY OF SECOND OBJECT BEFORE COLLISION)]VECTOR SUM OF
MOMENTUM BEFORE COLLISION $=$
[($m_{1,}$MASS OF FIRST OBJECT $v_{1f,}$ FINAL VELOCITY OF FIRST OBJECT AFTER COLLISION) $+$
($m_{2,}$MASS OF SECOND OBJECT $v_{2f,}$ FINAL VELOCITY OF SECOND OBJECT AFTER COLLISION)]VECTOR SUM OF MOMENTUM
AFTER COLLISION.

If you don't like all the subscripts, here it is, again! $m_1v_{1i} + m_2v_{2i} = m_1v_{1f} + m_2v_{2f}.$

$m_1 = m_{1,}$MASS OF FIRST OBJECT (BULLET), $v_{1i,}$ INITIAL VELOCITY OF BULLET BEFORE THE GUN IS FIRED $= 0$ m/s,
$m_{2,}$MASS OF SECOND OBJECT (GUN), $v_{2i,}$ INITIAL VELOCITY OF SECOND OBJECT (GUN) BEFORE IT IS FIRED $= 0.$

$\Rightarrow$ [$0 + 0$]MOMENTUM BEFORE THE GUN IS FIRED $=$
[($m_{1,}$MASS OF BULLET $v_{1f,}$ FINAL VELOCITY OF BULLET AFTER THE GUN IS FIRED) $+$
($m_{2,}$MASS OF GUN $v_{2f,}$ FINAL VELOCITY OF GUN AFTER IT IS FIRED)].

That is, the gun and the bullet have equal but opposite momentum so that the vector sum of the momentum is still zero even after we fire the gun!

Answer B!

17. By problem **16** arguments, we know that the bullet and the gun have equal but opposite momentum. Bullet mass is small and so we expect its velocity to be greater in magnitude than the velocity of the gun. Because kinetic energy terms have a square term involving the velocity, we expect its kinetic energy to be greater than the kinetic energy of the gun. Let us prove that mathematically.

If $P_{G,\text{ MOMENTUM OF GUN}}$ and $P_{B,\text{ MOMENTUM OF BULLET}}$ are momentum of the gun and bullet and $T_{G,}$ KINETIC ENERGY OF GUN and $T_{B,\text{ KINETIC ENERGY OF BULLET}}$ are their kinetic energies, then $P_{\text{GUN}} = -P_{\text{BULLET}}$ if momentum is conserved.

Kinetic energy of the gun is $T_{\text{GUN}} = P_{\text{GUN}}^2 / (2 \times M_{\text{GUN}})$.
Kinetic energy of the bullet is $T_{\text{BULLET}} = P_{\text{BULLET}}^2 / (2 \times M_{\text{BULLET}})$ where M_{GUN} and M_{BULLET} are the masses of the gun and the bullet respectively.

If $P_{\text{GUN}} = -P_{\text{BULLET}}$, then $[P_{\text{GUN}}]^2 = [-P_{\text{BULLET}}]^2$.
Note that the square of a negative number is positive.

Now we can rewrite the expressions of the kinetic energies in terms of momentum P_{GUN}.
That is:
$T_{\text{GUN}} = P_{\text{GUN}}^2 / (2 \times M_{\text{GUN}})$.   **(1)**
$T_{\text{BULLET}} = P_{\text{GUN}}^2 / (2 \times M_{\text{BULLET}})$.   **(2)**

Divide **(1)** by **(2)**.
$\Rightarrow T_{\text{GUN}} / T_{\text{BULLET}} = M_{\text{BULLET}} / M_{\text{GUN}}$.
Since mass of bullet is smaller, we expect the ratio $T_{\text{GUN}} / T_{\text{BULLET}}$ to be less than ONE. That can happen only if:
$T_{\text{KINETIC ENERGY OF BULLET}} >_{\text{GREATER THAN}} T_{\text{KINETIC ENERGY OF GUN}}$.
Answer C!

19. The direction of change in momentum *vector* determines the direction of the net force. Net force can be in any direction!

Problem **20** is a good example. Instead of the ball kicked to the left, assume it is kicked to the right with a speed v along $+x$. It then bounces off (elastically) against a wall and proceeds in the opposite direction along $-x$. $[\Delta v]_{\text{CHANGE IN VELOCITY}} = [v_f]_{\text{FINAL VELOCITY}} - [v_i]_{\text{INITIAL VELOCITY}}$.
$[\Delta v]_{\text{CHANGE IN VELOCITY}} = -v - v = -2v$. We hope you are very clear on this one.
Answer A!

20. $[\Delta v]_{\text{CHANGE IN VELOCITY}} = [v_f]_{\text{FINAL VELOCITY}} - [v_i]_{\text{INITIAL VELOCITY}}$.
Change in velocity $= (+v) - (-v) = 2v$.
Answer C!

22. Distance of the center of mass from m_2:

$= d_{\text{DISTANCE BETWEEN THE TWO MASSES}} - |x|_{\text{DISTANCE BETWEEN CENTER OF MASS AND MASS m1}} =$
$(m_{1,\text{ THE OTHER MASS}} \times d_{\text{DISTANCE BETWEEN TWO MASSES}}) \diagup (m_{1,\text{ MASS}} + m_{2,\text{ MASS}}).$
For simplicity, assume the origin is at the center of mass.
Answer A!

23. Distance of the center of mass of the *earth-moon* system from the center of earth =
$(m_{1,\text{ THE OTHER MASS}} \times d_{\text{DISTANCE BETWEEN THEIR CENTERS}}) \diagup (m_{1,\text{ MASS}} + m_{2,\text{ MASS}}).$

$\Rightarrow (m_{\text{MASS OF MOON}} \times 3.8 \times 10^8 \text{ m}) \diagup (m_{\text{MASS OF EARTH}} + m_{\text{MASS OF MOON}}) =$
$(m_{\text{MOON}} \times 3.8 \times 10^8) \diagup [(80 \times m_{\text{MOON}}) + m_{\text{MOON}})]$
$\approx (3.8 \times 10^8) \diagup 81$
$= 4.7 \times 10^6 \text{ m}.$

Remember that the mass of earth is 80 times the mass of the moon. It is stated in the problem.
The radius of the earth is 6.4×10^6.

The distance of the center of mass of the earth-moon system is only 4.7×10^6 m. That is the center of mass of the *earth-moon* system is very much within the radius of the earth (inside the earth) as 4.7×10^6 m is very much less than 3.8×10^8 m. Believe it!
Answer A!

25. This is an inelastic collision. Only the vector sum of momentum is conserved before and after the collision.
Use the famous equation:

$[(m_{1,\text{MASS OF FIRST OBJECT}} v_{1i,\text{ INITIAL VELOCITY OF FIRST OBJECT BEFORE COLLISION}}) +$
$(m_{2,\text{MASS OF SECOND OBJECT}} v_{2i,\text{ INITIAL VELOCITY OF SECOND OBJECT BEFORE COLLISION}})]_{\text{VECTOR SUM OF}}$
$_{\text{MOMENTUM BEFORE COLLISION}} =$
$[(m_{1,\text{MASS OF FIRST OBJECT}} v_{1f,\text{ FINAL VELOCITY OF FIRST OBJECT AFTER COLLISION}}) +$
$(m_{2,\text{MASS OF SECOND OBJECT}} v_{2f,\text{ FINAL VELOCITY OF SECOND OBJECT AFTER COLLISION}})]_{\text{VECTOR SUM OF MOMENTUM}}$
$_{\text{AFTER COLLISION}}.$

If you don't like all the subscripts, here it is, again! $m_1 v_{1i} + m_2 v_{2i} = m_1 v_{1f} + m_2 v_{2f}.$
$m_{1,\text{MASS OF FIRST OBJECT}} = \text{mass of bullet} = m.$
$v_{1i,\text{ INITIAL VELOCITY OF FIRST OBJECT BEFORE COLLISION}} = v_B.$
$m_{2,\text{MASS OF SECOND OBJECT}} = \text{mass of the wooden block.}$
$v_{2i,\text{ INITIAL VELOCITY OF SECOND OBJECT BEFORE COLLISION}} = 0 \text{ m/s.}$
$v_{1f,\text{ FINAL VELOCITY OF FIRST OBJECT AFTER COLLISION}} =$
$v_{2f,\text{ FINAL VELOCITY OF SECOND OBJECT AFTER COLLISION}} = v.$
Note that the velocity is the same for both the bullet and the block because they are traveling together as one unit.

By conservation of momentum, if v_B is the velocity of the bullet, then:

$[m_{\text{BULLET MASS}} \times v_{\text{VELOCITY OF BULLET}}]_{\text{VECTOR SUM OF MOMENTUM BEFORE COLLISION}} =$

$[(M_{\text{MASS OF BLOCK}} + m_{\text{MASS OF BULLET}}) \times v_{\text{VELOCITY OF THE TWO TOGETHER}}]_{\text{VECTOR SUM OF MOMENTUM AFTER COLLISION}}.$

$\Rightarrow v = (m \times v_B) / (M + m).$

Answer C!

26. The center of gravity of the block is raised through a height of h when the string is displaced through an angle θ given by: $h = l - (l \times cos\ \theta) = l \times (1 - cos\ \theta)$.

    Refer to problem **79** of the previous chapter.

    After the collision is complete, we can use energy conservation for the two masses. Just use it.

    Total Energy at the maximum height = Total Energy at minimum height.

    $[KE_{\text{KINETIC ENERGY}} + PE_{\text{POTENTIAL ENERGY}}]_{\text{TOTAL ENERGY AT MAXIMUM HEIGHT}} =$

    $[KE_{\text{KINETIC ENERGY}} + PE_{\text{POTENTIAL ENERGY}}]_{\text{TOTAL ENERGY AT MINIMUM HEIGHT}}.$

    $\Rightarrow 0_{\text{KINETIC ENERGY AT MAXIMUM HEIGHT}} + [(M + m)gh]_{\text{POTENTIAL ENERGY AT MAXIMUM HEIGHT}} =$

    $[(1/2) \times (m + M) \times v^2]_{\text{KINETIC ENERGY AT MINIMUM HEIGHT}} + [0]_{\text{POTENTIAL ENERGY AT MINIMUM HEIGHT}}.$

    Solving we get: $v^2 = 2 \times g \times h$.

    Substitute for $h = l \times (1 - cos\ \theta)$.

    $\Rightarrow v^2 = 2 \times g \times l \times (1 - cos\ \theta). \Rightarrow v = [2 \times g \times l \times (1 - cos\ \theta)]^{1/2}.$ **(1)**

    By momentum conservation (problem **25**), we have $mv_B = (M + m)\ v$. **(2)**

    Substitute for v from equation **(1)** in equation **(2)**.

    $\Rightarrow mv_B = (M + m) \times [2 \times g \times l \times (1 - cos\ \theta)]^{1/2}.$

    $\Rightarrow v_B = [(M + m) / m] \times [2 \times g \times l \times (1 - cos\ \theta)]^{1/2}.$

    Answer A!

28. Since kinetic energy is not conserved, the collision is inelastic. "Totally" inelastic collisions are characterized by the two bodies sticking together after the collision.

    Answer B!

29. Since force is the rate of change of momentum, the momentum is a constant in time when net force is zero for the system. If conservation of momentum applies, there are no external forces. So position and time should not matter!

    Answer A!

31. When potential energy is minimum, there is stable equilibrium.

    Answer B!

32. The potential energy of the body is minimum when its center of mass (center of gravity) is at the lowest position since potential energy = mass × *g* × height of center of mass above horizontal. Note that the center of mass and center of gravity are the same for point objects close to the surface of earth.
Answer A!

34. Vertical momentum at maximum height is zero. If we have totally elastic collision, then they bounce off each other. If on the other hand we have a totally inelastic collision, then they cancel momentum. Vertical momentum before they collide is zero. Horizontal momentum before and after collision is also zero as they are in opposite directions. This is as good as an object falling from rest! Wow! Wow! We love this question.
Answer D!

35. Magnitudes of the momentum of the two are the same if speed and mass are the same as in this problem, but not their directions. In a very strict sense, momentum is mass times velocity. It is not mass times speed! Again, momentum is a vector. Both magnitude and direction have to be the same for two vectors to be the same. Some students always forget to put a negative sign for one of the velocities when there is a head-on collision between two moving objects. Just do it right.
Answer B!

37. The beef of this problem is an important concept to understand. Remember that force is the rate of change of momentum. To get a bigger force, we need a large change in momentum for the same time interval. When velocities are reversed in direction (bouncing balls), one gets a larger change in momentum.
For example when we have a truly elastic collision, change in velocities =
$v_{\text{FINAL VELOCITY}} - (-v_{\text{INITIAL VELOCITY}}) = 2v$.

For the bullet that passes through, the change in momentum is nearly zero as incoming and outgoing speeds and directions are the same. There is little change in velocity. That is, force is nearly zero.
Answer A!

38. Before you throw the clothes, your momentum is zero. That means your vector sum of the momentum after you throw the clothes also should be zero. That means your clothes and you will have opposite directions.
Answer B!

40. $F_{\text{NET FORCE}} = m_{\text{MASS}} \times (\Delta v_{\text{CHANGE IN VELOCITY}} / \Delta t_{\text{TIME}}) =$
$\{0.05_{\text{MASS}} [(-60_{\text{FINAL VELOCITY}} - 60_{\text{INITIAL VELOCITY}})]\} / 0.02_{\text{TIME}} = 300 \text{ N along } -x$.
Remember that $\Delta v_{\text{CHANGE IN VELOCITY}} = v_{f, \text{ FINAL VELOCITY}} - v_{0, \text{ INITIAL VELOCITY}} = -60 - 60 = -120 \text{ m/s}$.
Answer B!

41.   We have to overcome gravity.

Gravitational force = $m_{MASS} \, g_{GRAV}$.

$\Rightarrow$ 300 N = $m \,(10)_{GRAV}$.

$\Rightarrow m_{MASS}$ = 30 kg.

Answer B!

43. $F_{NET\ FORCE} = (m_{MASS} \times \Delta v_{CHANGE\ IN\ VELOCITY}\,) / \Delta t_{TIME}$.

If net force is zero, the change in velocity has to be zero. Whenever there is a change of momentum, there has to be a non-zero net force acting on the object.

Answer C!

44.   1.   Use conservation of energy to find the speed of the fish just before it hits the net. That way we know the instantaneous initial speed with which the fish hits the net. The final speed of the fish is zero when the fish comes to rest.

2.   Find the change in momentum and divide it by time to get the force. The force applied by the net has to be opposite to that of the direction of fall for the fish. (The fish gets stopped. What do you think?) Force direction is along +y.

Equating total energy at the start = total energy at the end we get:

$(KE_{KINETIC\ ENERGY} + PE_{POTENTIAL\ ENERGY})_{AT\ THE\ START}$ =

$(KE_{KINETIC\ ENERGY} + PE_{POTENTIAL\ ENERGY})_{AT\ THE\ LANDING\ POINT}$.

$\Rightarrow [0]_{KINETIC\ ENERGY\ AT\ THE\ START} + [mgh_{HEIGHT\ OF\ FALL}]_{POTENTIAL\ ENERGY\ AT\ THE\ START}$ =

$[(1/2)mv_{SPEED}{}^2]_{KINETIC\ ENERGY\ AT\ THE\ LANDING\ POINT} + [0]_{POTENTIAL\ ENERGY\ AT\ ZERO\ LEVEL}$.

Or $mgh = (1/2)mv^2$.

$\Rightarrow v = (2gh)^{1/2}$.

$\Rightarrow v = (2 \times 10_{GRAV.} \times 2_{HEIGHT\ OF\ FALL})^{1/2}$.

$\Rightarrow v_{SPEED}$ = 6.32 m/s.

The fish is going down and so $v_{VELOCITY}$ = – 6.32 m/s is more correct.

Change in velocity for the fish after hitting the net =

$0_{FINAL\ VELOCITY} - (-6.32)_{INITIAL\ VELOCITY}$ = +6.32 m/s.

$\Rightarrow$ Change in momentum = $m_{MASS} \times \Delta v_{CHANGE\ IN\ VELOCITY}$ = 2 kg × (+6.32 m/s) = 12.64 m/s.

(Average) Force = $(m_{MASS} \times \Delta v_{CHANGE\ IN\ VELOCITY}) / t_{TIME}$ = 12.64 / 0.7 = 18.05 N along +y.

Answer C!

46.   To create a net force, one has to change the momentum. That is, whenever we want to change the momentum, a non-zero force has to be applied for a period of time. Impulse = $F_{FORCE} \times t_{TIME}$. This is nothing but change in momentum, $m_{MASS}\Delta v_{CHANGE\ IN\ VELOCITY}$. Do you want it one more time?

Impulse = $F_{FORCE} \times t_{TIME} = m_{MASS}\Delta v_{CHANGE\ IN\ VELOCITY}$.

Answer D!

47.   Impulse = $F_{FORCE} \times t_{TIME} = m_{MASS}\Delta v_{CHANGE\ IN\ VELOCITY}$.

⇒ $F_{FORCE} = [m_{MASS}\Delta v_{CHANGE\ IN\ VELOCITY}] / t_{TIME}$.

Our goal is to do less harm to ourselves. If we use the language of physics, we need to reduce the force.

$F_{FORCE} = [m_{MASS}\Delta v_{CHANGE\ IN\ VELOCITY}] / t_{TIME}$.

The only way to reduce the force during a collision is to increase the time of contact.

Answer B!

49. $F_{FORCE} = [m_{MASS}\Delta v_{CHANGE\ IN\ VELOCITY}] / t_{TIME}$.

Your goal here is to reduce the force. So you increase the time of contact.

Answer D!

50. $F_{FORCE} = [m_{MASS}\Delta v_{CHANGE\ IN\ VELOCITY}] / t_{TIME}$.

Here your goal is to get points by boxing within the rules, and not with bare fists. To decrease the force of impact, you have to increase the time of contact. Padding helps to increase the time of contact!

Answer A!

52.   Find the speed just before landing. At the landing point, total energy is kinetic. Do backwards and find the height of fall.

STEP 1: $(m_{MASS} \times \Delta v_{CHANGE\ IN\ VELOCITY}) / t_{TIME}$ = Force =

$70_{MASS} \times (0_{FINAL\ VELOCITY} - v_{0,\ INITIAL\ VELOCITY}) / 0.04_{TIME} = 20{,}000$.

⇒ $v_0 = -11.43$ m/s.

It is negative because the initial velocity is vertically down assuming **+y** is vertically upwards. Once we know the velocity (actually speed), we can use energy conservation to find the height.

STEP 2:   In the absence of friction and input energy, total energy is conserved.

$(KE_{KINETIC\ ENERGY} + PE_{POTENTIAL\ ENERGY})_{AT\ THE\ START} =$
$(KE_{KINETIC\ ENERGY} + PE_{POTENTIAL\ ENERGY})_{AT\ THE\ LANDING\ POINT}$.

⇒ $[0]_{KINETIC\ ENERGY\ AT\ THE\ START} + [mgh_{HEIGHT\ OF\ FALL}]_{POTENTIAL\ ENERGY\ AT\ THE\ START} =$
$[(1/2)mv_{SPEED}^2]_{KINETIC\ ENERGY\ AT\ THE\ LANDING\ POINT} + [0]_{POTENTIAL\ ENERGY\ AT\ ZERO\ LEVEL}$.

Or $mgh = (1/2)mv^2$. ⇒ $h = (v^2 / 2g)$.

⇒ $h = (11.43_{VELOCITY})^2 / (2 \times 10_{GRAV.})$.

⇒ $h_{HEIGHT\ OF\ FALL} = 6.5$ m.

QUICK SOLUTION: Set Total Energy at the beginning = Total Energy at the end.

$(KE + PE)_{BEGINNING} = (KE + PE)_{END}$. ⇒ $0 + mgh = (1/2)mv^2 + 0$. ⇒ $h = [(1/2)v^2] / g = 6.5$ m if $v = 11.43$ m/s.

Answer D!

53. m_1 is moving to the right at constant velocity and so net force on m_1 before collision is zero.

Remember that $F_{FORCE} = [m_{MASS}\Delta v_{CHANGE\ IN\ VELOCITY}] / t_{TIME}$.

If change in velocity is zero, force is zero!

Answer A!

55. Since m_1 moves to the right and hits m_2, net force on mass m_1 is to the left due to mass m_2. Similarly net force on mass m_2 is to the right due to mass m_1. [Note: If you consider masses m_1 and m_2 together as the system, these forces are equal and opposite. So they cancel, giving a net force of zero on the system consisting of two colliding masses at the instant they collide.]

Answer C!

56. Net force on mass m_2 due to mass m_1 is to the right if mass m_2 is to the right of mass m_1.

Answer B!

58. Use the famous equation:

$[(m_{1,MASS\ OF\ FIRST\ OBJECT}\ v_{1i,\ INITIAL\ VELOCITY\ OF\ FIRST\ OBJECT\ BEFORE\ COLLISION}) +$

$(m_{2,MASS\ OF\ SECOND\ OBJECT}\ v_{2i,\ INITIAL\ VELOCITY\ OF\ SECOND\ OBJECT\ BEFORE\ COLLISION})]_{VECTOR\ SUM\ OF}$

MOMENTUM BEFORE COLLISION $=$

$[(m_{1,MASS\ OF\ FIRST\ OBJECT}\ v_{1f,\ FINAL\ VELOCITY\ OF\ FIRST\ OBJECT\ AFTER\ COLLISION}) +$

$(m_{2,MASS\ OF\ SECOND\ OBJECT}\ v_{2f,\ FINAL\ VELOCITY\ OF\ SECOND\ OBJECT\ AFTER\ COLLISION})]_{VECTOR\ SUM\ OF\ MOMENTUM}$

AFTER COLLISION.

If you don't like all the subscripts, here it is, again! $m_1 v_{1i} + m_2 v_{2i} = m_1 v_{1f} + m_2 v_{2f}$.

$m_{1,MASS\ OF\ FIRST\ OBJECT} = $ mass of car $= 10{,}000$ kg.

$v_{1i,\ INITIAL\ VELOCITY\ OF\ FIRST\ OBJECT\ BEFORE\ COLLISION} = 12$ m/s.

$m_{2,MASS\ OF\ SECOND\ OBJECT} = $ mass of the other car $=10{,}000$ kg.

$v_{2i,\ INITIAL\ VELOCITY\ OF\ SECOND\ OBJECT\ BEFORE\ COLLISION} = 0$.

$v_{1f,\ FINAL\ VELOCITY\ OF\ FIRST\ OBJECT\ AFTER\ COLLISION} = $

$v_{2f,\ FINAL\ VELOCITY\ OF\ SECOND\ OBJECT\ AFTER\ COLLISION} = v$.

By conservation of momentum, $m_1 v_{1i} + m_2 v_{2i} = m_1 v_{1f} + m_2 v_{2f}$.

$\Rightarrow (10{,}000_{MASS} \times 12_{VELOCITY})_{FIRST\ CAR} + (10{,}000_{MASS} \times 0_{VELOCITY})_{SECOND\ CAR} =$
$(10{,}000_{MASS} + 10{,}000_{MASS})_{TOTAL\ MASS} \times v_{VELOCITY\ OF\ THE\ TWO\ CARS\ AFTER\ COLLISION}$.

$\Rightarrow v_{VELOCITY\ OF\ THE\ TWO\ CARS\ TOGETHER\ AFTER\ COLLISION} = 6$ m/s.

Answer C!

59. $[(m_{1,\text{MASS OF FIRST OBJECT}}\, v_{1i},\ _{\text{INITIAL VELOCITY OF FIRST OBJECT BEFORE COLLISION}}) +$

$(m_{2,\text{MASS OF SECOND OBJECT}}\, v_{2i},\ _{\text{INITIAL VELOCITY OF SECOND OBJECT BEFORE COLLISION}})]_{\text{VECTOR SUM OF}}$

MOMENTUM BEFORE COLLISION $=$

$[(m_{1,\text{MASS OF FIRST OBJECT}}\, v_{1f},\ _{\text{FINAL VELOCITY OF FIRST OBJECT AFTER COLLISION}}) +$

$(m_{2,\text{MASS OF SECOND OBJECT}}\, v_{2f},\ _{\text{FINAL VELOCITY OF SECOND OBJECT AFTER COLLISION}})]_{\text{VECTOR SUM OF MOMENTUM}}$

AFTER COLLISION.

If you don't like all the subscripts, here it is, again! $m_1 v_{1i} + m_2 v_{2i} = m_1 v_{1f} + m_2 v_{2f}.$

$\qquad m_{1,\text{MASS OF FIRST OBJECT}} = $ mass of car $= 10{,}000$ kg.

$\qquad v_{1i},\ _{\text{INITIAL VELOCITY OF FIRST OBJECT BEFORE COLLISION}} = 12$ m/s.

$\qquad m_{2,\text{MASS OF SECOND OBJECT}} = $ mass of the other car $= 10{,}000$ kg.

$\qquad v_{2i},\ _{\text{INITIAL VELOCITY OF SECOND OBJECT BEFORE COLLISION}} = -12$ m/s.

$\qquad v_{1f},\ _{\text{FINAL VELOCITY OF FIRST OBJECT AFTER COLLISION}} =$

$\qquad\qquad v_{2f},\ _{\text{FINAL VELOCITY OF SECOND OBJECT AFTER COLLISION}} = v.$

By conservation of momentum, $m_1 v_{1i} + m_2 v_{2i} = m_1 v_{1f} + m_2 v_{2f}.$

$\Rightarrow (10{,}000_{\text{MASS}} \times 12_{\text{VELOCITY}})_{\text{FIRST CAR}} + (10{,}000_{\text{MASS}} \times -12_{\text{VELOCITY}})_{\text{SECOND CAR}} =$
$(10{,}000_{\text{MASS}} + 10{,}000_{\text{MASS}})_{\text{TOTAL MASS}} \times v_{\text{VELOCITY OF THE TWO CARS AFTER COLLISION}}.$

By conservation of momentum, $m_1 v_{1i} + m_2 v_{2i} = m_1 v_{1f} + m_2 v_{2f}.$

$(10{,}000 \times 12) + (10{,}000 \times -12) = (10{,}000 + 10{,}000) \times v.$

$\Rightarrow v = 0$ m/s.

Note that one of the signs of the velocities is reversed. Are you surprised? I was when I was an undergraduate student! Long time ago!

Answer E!

61. Note that we can use equations of kinematics or energy conservation to find the speed at the bottom just before impact. Let us go with equations of kinematics. It will be a good review for you.

Any object kept on an inclined plane in the absence of friction will have:

$[m_{\text{MASS}}\, g_{\text{GRAV.}} \times \sin\theta_{\text{ANGLE OF THE INCLINED PLANE}}]_{\text{THE ONLY FORCE TOWARDS THE BASE}} =$
$[m_{\text{MASS}} a_{\text{ACCELERATION}}]_{\text{NET FORCE}}.$

$\Rightarrow a_{\text{ACCELERATION}} = $ acceleration along the inclined plane $= [g \times \sin\theta_{\text{ANGLE OF THE INCLINED PLANE}}].$

Now we can use one of the equations of kinematics for constant acceleration to calculate the speed at the bottom.

Use $[v_{\text{FINAL SPEED}}^2] = [v_{0,\text{ INITIAL SPEED}}^2] + [2a_{\text{ACCELERATION}}\, \Delta x_{\text{DISTANCE}}].$

$\Rightarrow v^2 = 0^2 + (2g \times \sin 30 \times 12_{\text{DISTANCE}}).$

$\Rightarrow v_{\text{SPEED}} = 10.95$ m/s.

Answer A!

62. By momentum conservation: $m_1v_{1i} + m_2v_{2i} = m_1v_{1f} + m_2v_{2f}$.

$[(2_{MASS} \times 10.95_{VELOCITY}) + (4_{MASS} \times 0_{VELOCITY})]_{VECTOR\ SUM\ OF\ MOMENTUM\ BEFORE\ COLLISION} =$

$[(2_{MASS} + 4_{MASS}) \times v_{VELOCITY}.]_{VECTOR\ SUM\ OF\ MOMENTUM\ AFTER\ COLLISION}.$

$\Rightarrow v_{VELOCITY} = 3.65$ m/s.

Answer A!

64. By momentum conservation: $m_1v_{1i} + m_2v_{2i} = m_1v_{1f} + m_2v_{2f}$.

$(2_{MASS} \times 10.95_{VELOCITY}) + (4_{MASS} \times -6_{VELOCITY}) = (2_{MASS} + 4_{MASS}) \times v_{VELOCITY}.$ $\Rightarrow v = -0.35$ m/s (to the left).

Answer E!

65. Find the speed at the bottom.

Set Total Energy at the bottom = Total Energy at the top to get the total height.

By momentum conservation: $m_1v_{1i} + m_2v_{2i} = m_1v_{1f} + m_2v_{2f}$.

$(2_{MASS} \times 10.95_{VELOCITY}) + (4_{MASS} \times -6_{VELOCITY}) = (2_{MASS} + 4_{MASS}) \times v_{VELOCITY}.$

$\Rightarrow v = -0.35$ m/s to the left.

$TE_{TOTAL\ ENERGY}$ at the bottom = $TE_{TOTAL\ ENERGY}$ at the top.

$\Rightarrow (KE + PE)_{BOTTOM} = (KE + PE)_{TOP}.$

$\Rightarrow [(1/2)mv^2]_{KINETIC\ ENERGY\ AT\ THE\ BASE} + [0]_{POTENTIAL\ ENERGY\ AT\ THE\ BASE} =$

$[0]_{KINETIC\ ENERGY\ AFTER\ IT\ COMES\ TO\ A\ STOP} + [mgh]_{POTENTIAL\ ENERGY\ AT\ THE\ TOP}.$

$\Rightarrow h = [(1/2)v^2] / g = 0.0061$ m.

Answer A!

67. By momentum conservation: $m_1v_{1i} + m_2v_{2i} = m_1v_{1f} + m_2v_{2f}$.

$(3_{MASS} \times 5_{VELOCITY}) + (2_{MASS} \times 0_{VELOCITY}) = (3_{MASS} + 2_{MASS}) \times v_{VELOCITY}.$

$\Rightarrow v_{HORIZONTAL\ VELOCITY\ AFTER\ COLLISION} = 3$ m/s.

Collision takes place on a horizontal table. Because velocity does not have a vertical component at the start, initial momentum along *y* is zero. By momentum conservation, final momentum of the two masses after collision should also be zero. This makes the vertical instantaneous speed just after collision, zero.

Answer A!

68. The projectile after collision is shooting off horizontally. That is, the (initial) vertical velocity immediately after collision is zero.

Use the third equation of kinematics: $\Delta y = [v_{0y}t] + [(1/2)a_yt^2]$.

If **+y** is upwards, $\Delta y = -4$ m (final point is below the starting point), $v_{0y} = 0$ m/s, $a_y = -10$ m/s/s then;

$t = [(2\Delta y) / g]^{1/2} = [(2 \times -4) / -10]^{1/2} = 0.89$ s.

Answer B!

70. By momentum conservation: $m_1v_{1i} + m_2v_{2i} = m_1v_{1f} + m_2v_{2f}$.

$(3_{MASS} \times 5_{VELOCITY}) + (2_{MASS} \times 0_{VELOCITY}) = (3_{MASS} + 2_{MASS}) \times v_{HORIZONTAL\ VELOCITY\ AFTER\ COLLISION}$.

$\Rightarrow v = 3$ m/s.

Use the third equation of kinematics: $\Delta y = v_{0y}t + (1/2)a_yt^2$.

Please look at solution to problem **68**.

If +y is upwards, $\Delta y = -4$ m, $v_{0y} = 0$ m/s, $a_y = -10$ m/s/s; then $t = 0.89$ s.

Horizontal displacement $= \Delta x = [v_{0x}]_{HORIZONTAL\ VELOCITY} \times [t]_{TIME}$.

$\Rightarrow \Delta x_{HORIZONTAL\ DISPLACEMENT} = [3\ m/s]_{HORIZONTAL\ VELOCITY} \times [0.89\ s]_{TIME} = 2.67$ m.

Answer C!

71. At the instant they collide, momentum is conserved. Because the collision is horizontal, vertical velocity before and after collision is the same, 0! After collision, gravitational force makes the vertical component of the velocity increase in magnitude.

Answer A!

73. Gravitational force acts along the vertical.

Answer B!

74. Impulse = force × time = $[(m_1 + m_2)g]_{GRAVITATIONAL\ FORCE} \times t_{TIME}$.

Answer C!

76. By momentum conservation: $m_1v_{1i} + m_2v_{2i} = m_1v_{1f} + m_2v_{2f}$.

$(0.2_{MASS} \times 400_{VELOCITY}) + (2_{MASS} \times -2_{VELOCITY}) = (0.2_{MASS} + 2_{MASS}) \times v_{VELOCITY\ AFTER\ COLLISION}$. $\Rightarrow v = 34.55$ m/s.

Answer C!

77. By momentum conservation: $m_1v_{1i} + m_2v_{2i} = m_1v_{1f} + m_2v_{2f}$.

$(0.2_{MASS} \times 400_{VELOCITY}) + (2_{MASS} \times +2_{VELOCITY}) = (0.2_{MASS} + 2_{MASS}) \times v_{VELOCITY\ AFTER\ COLLISION}$. $\Rightarrow v = 38.2$ m/s.

Answer B!

1. Torque = F VERTICAL FORCE $\times R$ PERPENDICULAR (HORIZONTAL) DISTANCE clockwise.
    Note that force can take any direction. (If force acts along the vertical, then distance perpendicular to the line of action of the force is horizontal. Never make a mistake on that concept.) It is going to rotate!
    Answer A!

2. Force makes the lever arm rotate counter-clockwise.
    $\sum \tau$NET TORQUE $= I$ ROTATIONAL INERTIA $\times \alpha$ ANGULAR ACCELERATION
    $= F$ VERTICAL FORCE $\times R$ PERPENDICULAR (HORIZONTAL) DISTANCE counterclockwise about an axis passing
    through point A. The moment of inertia, I, is usually not zero if the mass is not zero. $\Rightarrow \alpha \neq 0$ or it will pick up angular speed!
    Answer B!

4. No matter how hard you push on the lever arm, it will *not* rotate about an axis passing through point A. You can say (correctly) that the line of action of this horizontal force passes through the axis of rotation.
    $\sum \tau$NET TORQUE $= I$ ROTATIONAL INERTIA $\times \alpha$ ANGULAR ACCELERATION
    $= F$ HORIZONTAL FORCE $\times (R \times sin\ 0)$ PERPENDICULAR (VERTICAL) DISTANCE $= 0$.
    $\Rightarrow I \times \alpha = 0.\ \Rightarrow \alpha = 0.$
    Answer C!

5. $(F \times cos\ 30)$ HORIZONTAL COMPONENT OF THE FORCE will not rotate the lever arm about an axis passing through point A because the lever arm is zero for that component. Or you can say (correctly) that the line of action of the horizontal component of the force passes through the axis of rotation.
    But $(F \times sin\ 30)$VERTICAL COMPONENT OF THE FORCE will produce an unbalanced torque.
    τ NET TORQUE $= [(F \times cos\ 30)$HORIZONTAL COMPONENT OF THE FORCE $\times 0$ PERPENDICULAR DISTANCE TO THE AXIS OF ROTATION] ZERO TORQUE $+$
    $[(F \times sin\ 30)$VERTICAL COMPONENT OF THE FORCE $\times R$ PERPENDICULAR (HORIZONTAL) DISTANCE $)]$ CLOCKWISE TORQUE clockwise.
    $\sum \tau$NET TORQUE $= I$ ROTATIONAL INERTIA $\times \alpha$ ANGULAR ACCELERATION $= F \times sin\ 30 \times R$ clockwise $= I \times \alpha$.
    $\Rightarrow \alpha$ is also clockwise.
    Answer D!

7. No matter how much force you apply at the hinge, a door will not rotate BECAUSE the line of action of the force passes through the axis of rotation. You must apply the force at a point away from the hinge to produce a torque.
    $\sum \tau$NET TORQUE $= I$ ROTATIONAL INERTIA $\times \alpha$ ANGULAR ACCELERATION $= 0 =$
    F VERTICAL FORCE $\times 0$ PERPENDICULAR (HORIZONTAL) DISTANCE TO THE AXIS OF ROTATION. $\Rightarrow I \times \alpha = 0.\ \Rightarrow \alpha = 0.$
    Answer C!

8. No rotation. No matter how much force you apply at the hinge, a door will not rotate BECAUSE the line of action of the vertical force passes through the axis of rotation. You must apply the force at a point away from the hinge to produce a torque.

$\sum_T$ NET TORQUE = I ROTATIONAL INERTIA $\times \alpha$ ANGULAR ACCELERATION = 0 =
F VERTICAL FORCE $\times 0$ PERPENDICULAR (HORIZONTAL) DISTANCE TO THE AXIS OF ROTATION·
$\Rightarrow I \times \alpha = 0. \Rightarrow \alpha = 0.$
Answer C!

10. 1. F_1 is acting at the axis of rotation, so it will not contribute to torque BECAUSE the line of action of this horizontal force passes through the axis of rotation.

2. $(F \times \cos 30)$ HORIZONTAL COMPONENT OF THE FORCE is a pulling force along the rod so will not contribute to torque either.

3. $(F \times \sin 30)$ VERTICAL COMPONENT OF THE FORCE is perpendicular to the lever arm, so we get a net torque in the counter-clockwise.

$\sum_T$ NET TORQUE = I ROTATIONAL INERTIA $\times \alpha$ ANGULAR ACCELERATION =
$[(F \times \sin 30)$ VERTICAL COMPONENT OF THE FORCE $\times R$ PERPENDICULAR (HORIZONTAL) DISTANCE TO THE AXIS OF ROTATION]COUNTERCLOCKWISE = $I \times \alpha$ counterclockwise about the axis of rotation.
Answer E!

11. 1.   Torque of the upper force causes the object to rotate in the clockwise direction.

2.   Torque of the lower force causes the object to rotate counterclockwise.

These are equal and opposite torques. $\Rightarrow \tau_{net} = 0 = I \times \alpha.$
$\Rightarrow \alpha = 0.$

<u>Here are the details:</u>

1.   Torque of the upper force

= F HORIZONTAL FORCE $\times (R / 2)$ PERPENDICULAR (VERTICAL) DISTANCE TO THE AXIS OF ROTATION clockwise.

2.   Torque of the lower force

= F HORIZONTAL FORCE $\times (R / 2)$ PERPENDICULAR (VERTICAL) DISTANCE TO THE AXIS OF ROTATION counterclockwise.

These torques are equal and opposite.

$\Rightarrow \sum_T$ NET TORQUE = I ROTATIONAL INERTIA $\times \alpha$ ANGULAR ACCELERATION = 0.
$\Rightarrow I \times \alpha = 0. \Rightarrow \alpha = 0.$
Answer C!

13. 1.   Torque of the upper force =

F HORIZONTAL FORCE $\times (R / 2)$ PERPENDICULAR (VERTICAL) DISTANCE TO THE AXIS OF ROTATION counterclockwise.

2.   Torque of the lower force =

F HORIZONTAL FORCE $\times (R / 2)$ PERPENDICULAR (VERTICAL) DISTANCE TO THE AXIS OF ROTATION clockwise.

These torques are equal and opposite.

$\Rightarrow \sum_T$ NET TORQUE = I ROTATIONAL INERTIA $\times \alpha$ ANGULAR ACCELERATION = 0.
$\Rightarrow I \times \alpha = 0. \Rightarrow \alpha = 0.$
Answer C!

14. 1. No matter how much force you apply at the axis of rotation, the object will still not rotate. F is acting at the axis of rotation, so it will not contribute to torque BECAUSE the line of action of this horizontal force passes through the axis of rotation. Torque for this force $F = 0$.

2. Torque of the upper force

$= F$ HORIZONTAL FORCE $\times (R / 2)$PERPENDICULAR (VERTICAL) DISTANCE TO THE AXIS OF ROTATION clockwise

$= F \times (R / 2)$ clockwise.

3. Torque of the lower force

$= F$ HORIZONTAL FORCE $\times (R / 2)$PERPENDICULAR (VERTICAL) DISTANCE TO THE AXIS OF ROTATION clockwise

$= F \times (R / 2)$ also clockwise.

Torques (2. and 3.) are equal and in the *same* direction. They add!

$\Rightarrow \sum_T$ NET TORQUE $= I$ ROTATIONAL INERTIA $\times \alpha$ ANGULAR ACCELERATION $\neq 0$

$= [F \times (R / 2)]$ CLOCKWISE $+ [F \times (R / 2)]$ CLOCKWISE $= F \times R$ clockwise. $\Rightarrow I \times \alpha \neq 0$. $\Rightarrow \alpha \neq 0$.

Answer A!

16. For the situation shown, net force is along $+ y$.

Answer A!

17. Along $+ x$.

Answer C!

19. Net force has $-x$ and $-y$ components as $\sum F_x \neq 0$ and $\sum F_y \neq 0$.

Answer E!

20. Net force has $+x$ and $+y$ components. $\sum F_x = F \times cos\ 30$ and $\sum F_y = F \times sin\ 30$.

Answer E!

22. $\sum F_y = F$ along $+ y$.

Answer A!

23. $\sum F_x = F_1$ along $+ x$.

Answer C!

25. $\sum F_x = F + F$ along $+x$.

Answer C!

26. $\sum F_x = F + (- F) + (- F)$ to the left along $-x$.

Answer D!

28. $\sum F_x = F + F + (- F) = F$ along $+ x$ to the right.

Answer C!

29.   Redraw the figure with x and y force components.

$[\sum F_x]$ VECTOR SUM OF THE HORIZONTAL FORCES $= 0$ gives us:

$[F_3 \times cos\ 60]$ HORIZONTAL COMPONENT OF THE FORCE TO THE LEFT $= F_1$, HORIZONTAL FORCE TO THE RIGHT.

Answer D!

31.   Identify the horizontal and vertical forces or components of the forces first. Let us take the torque of each of those forces.

a.   If F_1 alone is acting at point A, the torque due to F_1 acting at the axis of rotation will be zero BECAUSE the line of action of this horizontal force passes through the axis of rotation.

b.   If F_2 alone is acting about point A, then F_2 will make the object rotate clockwise.

Now we have to decide on the perpendicular distance to force F_2.

The distance perpendicular to F_2 is 3.5 meters.

$\Rightarrow$ Torque due to F_2

$= [(F_2)$ VERTICAL FORCE $\times (\ 3.5)$ PERPENDICULAR (HORIZONTAL) DISTANCE TO THE AXIS OF ROTATION

clockwise. Note that $3.5 = 7/2$!

c.   If $F_3 \times cos\ 60$ alone is acting, then the torque $= F_3 \times cos\ 60 \times 3$ is counterclockwise.

$\Rightarrow$ $[(F_3 \times cos\ 60)$ HORIZONTAL COMPONENT OF THE FORCE $\times 3$ PERPENDICULAR (VERTICAL) DISTANCE TO THE

AXIS OF ROTATION $]$ counterclockwise.

d.   If $F_3 \times sin\ 60$ alone is acting about A, it will not cause the object to rotate.

$\tau = 0$. BECAUSE the line of action of this vertical force passes through the axis of rotation!

Now let us put the sum of all the clockwise torques equal to the sum of all the counterclockwise torques.

This gives us: $[F_2 \times 3.5]$ SUM OF THE CLOCKWISE TORQUES ABOUT THE AXIS OF ROTATION

$= [(F_3 \times cos\ 60) \times 3]$ SUM OF THE COUNTERCLOCKWISE TORQUES ABOUT THE AXIS OF ROTATION.

Answer C!

32.   To make the glass stand upright, we need to rotate it clockwise against the torque of its weight acting at the center of gravity.

Answer C!

34.   The forces acting along y are the normal force N, and the weight, $m \times g$.

$\Rightarrow N$ NORMAL FORCE $= m$ MASS $\times g$ GRAV. if we need equilibrium.

$[N$ NORMAL FORCE $]$ VECTOR SUM OF THE FORCES VERTICALLY UP

$= [m$ MASS $\times g$ GRAV. $]$VECTOR SUM VERTICALLY DOWN.

Answer B!

35.   Identify the horizontal and vertical forces or components of the forces first. Let us take the torque of each of those forces.

a.   If weight alone acts about point A, the box will have a torque in the counterclockwise direction.

Torque of weight about A $= m \times g \times 2.5$ counterclockwise.

Note that 2.5 m is the perpendicular distance to the axis of rotation.

Torque of weight about A $=$

[$(m \times g)$ VERTICAL FORCE $\times (2.5)$PERPENDICULAR (HORIZONTAL) DISTANCE TO THE AXIS OF ROTATION] counterclockwise.

b.   If F_{push} alone acts on the box about point A, then F_{push} will give a torque in the clockwise direction with a perpendicular distance of 3.5 m.

Torque of F_{push} about A = $F_{push} \times 3.5$ clockwise.

Torque of F_{push} about A =

[(F_{push}) HORIZONTAL FORCE $\times (3.5)$ PERPENDICULAR (VERTICAL) DISTANCE TO THE AXIS OF ROTATION] clockwise.

c.   Torque of the normal force is zero as it is acting at the axis of rotation.

Now add all the clockwise torques and set them equal to the sum of the counterclockwise torques:

[$m \times g \times 2.5$] TOTAL SUM OF COUNTERCLOCKWISE TORQUES ABOUT THE AXIS OF ROTATION =

[$F_{push} \times 3.5$]TOTAL SUM OF CLOCKWISE TORQUES ABOUT THE AXIS OF ROTATION.

Answer E!

37. $\sum F_y$ = The vector sum of the vertical forces = 0.

$\sum F_y$ VECTOR SUM OF THE FORCES ALONG THE VERTICAL $= 0$.

Or we get: [$F_1 + F_2$] THE SUM OF THE FORCES ACTING VERTICALLY UP =

[$m \times g$]THE SUM OF THE FORCES ACTING VERTICALLY DOWN.

Answer C!

38.   [$\sum F_y$ VECTOR SUM OF THE FORCES ALONG THE VERTICAL $= 0$.] $\sum F_y = 0$. $\Rightarrow [N_R + N_F] = [m \times g]$.

With details we get:

[N_R, NORMAL FORCE AT THE REAR $+ N_F$, NORMAL FORCE AT THE FRONT] = [m MASS $\times g$ GRAV.].

Or we get: [$N_R + N_F$] THE SUM OF THE FORCES ACTING VERTICALLY UP =

[$m \times g$]THE SUM OF THE FORCES ACTING VERTICALLY DOWN.

Answer B!

40.   [$\sum F_y$ THE VECTOR SUM OF THE FORCES ALONG THE VERTICAL $= 0$.] $\Rightarrow [W_2 + (m \times g)] = [N_R + N_F]$.

With details:

[W_2, WEIGHT OF THE GROCERIES $+ (m \times g)$WEIGHT OF THE CAR] SUM OF THE FORCES ACTING VERTICALLY DOWN =

[N_R, NORMAL FORCE AT THE REAR $+ N_F$, NORMAL FORCE AT THE FRONT] SUM OF THE FORCES ACTING VERTICALLY UP.

Answer C!

41.   Identify the horizontal and vertical forces or components of the forces first. Let us take the torque of each of those forces.

1.   Torque of W_2 about point A = $W_2 \times 2$ counter-clockwise.

With details we can write:

[W_2, WEIGHT OF GROCERIES (VERTICAL FORCE) $\times 2$ PERPENDICULAR (HORIZONTAL) DISTANCE TO THE AXIS OF ROTATION] counterclockwise.

2.   Torque of N_R about point A = [N_R, NORMAL FORCE AT THE REAR $\times 0$] = 0! BECAUSE the line of action

of this vertical force passes through the axis of rotation.

3.   Torque of $m \times g$ about point A = $[(m \times g) \times 0.75]$ clockwise. With details we can write:

$[(m \times g)$ WEIGHT OF THE CAR (VERTICAL FORCE) $\times 0.75$ PERPENDICULAR (HORIZONTAL) DISTANCE TO THE AXIS OF ROTATION] clockwise.

4.   Torque of N_F about point A = $N_F \times (3 + 0.75)$ counter-clockwise.

With details, we can write:

$[N_F,$ NORMAL FORCE AT THE FRONT (VERTICAL FORCE) $\times (3 + 0.75)$ PERPENDICULAR (HORIZONTAL) DISTANCE TO THE AXIS OF ROTATION] counter-clockwise.

Net torque about the axis of rotation, $\sum \tau = 0$.

$\Rightarrow$ [Sum of clockwise torques = Sum of counterclockwise torques] ABOUT THE AXIS OF ROTATION.

$\Rightarrow [m \times g \times 0.75]$ SUM OF THE CLOCKWISE TORQUES $=$

$[(W_2 \times 2) + (N_F \times 3.75)]$ SUM OF THE COUNTER-CLOCKWISE TORQUES.

Answer A!

43.   The vector sum of the forces along the horizontal $\sum F_x = 0$.

$\Rightarrow [F_H] = [T]$.

With details we can write:

$[F_H,$ HORIZONTAL FORCE TO THE RIGHT AT THE SUPPORT] $=$

$[T$ TENSION WITHIN THE STRING TO THE LEFT ACTING ON THE BEAM].

Answer C!

44.   The vector sum of the forces along the vertical $\sum F_y = 0$.

$\Rightarrow F_V = (m_{\text{beam}} \times g) + (m_1 \times g)$.

With details we can write:

$[F_V]$ SUM OF THE FORCES ACTING VERTICALLY UP $=$

$[(m_{\text{beam}} \times g)$ WEIGHT OF BEAM $+ (m_1 \times g)$ WEIGHT OF SOMETHING ELSE] SUM OF THE FORCES ACTING VERTICALLY DOWN.

Answer B!

46.   The vector sum of the forces along the horizontal acting on the beam $\sum F_x = 0$.

$\Rightarrow T \times \cos 30 = F_H$.

With details we can write:

$[T \times \cos 30]$ HORIZONTAL COMPONENT OF TENSION TO THE LEFT $=$

$[F_H,$ HORIZONTAL FORCE AT THE SUPPORT ACTING ON THE BEAM TO THE RIGHT].

Answer A!

47.   The vector sum of the forces along the vertical acting on the beam $\sum F_y = 0$.

$\Rightarrow [F_V$ FORCE ACTING VERTICALLY UP $+ (T \times \sin 30)$ VERTICAL COMPONENT OF TENSION ACTING VERTICALLY UP] $=$

$[(m_1 \times g) + (m_2 \times g)]$ SUM OF THE FORCES ACTING VERTICALLY DOWN.

Answer A!

49.   1.   Torque of $(m_2 \times g)$ about an axis passing through point A =

[$(m_2 \times g)$ VERTICAL FORCE $\times 1$ PERPENDICULAR (HORIZONTAL) DISTANCE TO THE AXIS OF ROTATION]
counterclockwise.

2.   Torque of $(m_1 \times g)$ about an axis passing through point A =

[$(m_1 \times g)$ VERTICAL FORCE $\times 3$ PERPENDICULAR (HORIZONTAL) DISTANCE TO THE AXIS OF ROTATION]
clockwise.

3.   Torque of support force at A = 0.

BECAUSE the line of action of this vertical force passes through the axis of rotation.

Net torque about the axis of rotation $\sum \tau = 0.$ $\Rightarrow$ Sum of clockwise torques about the axis of rotation = Sum
of counterclockwise torques about the axis of rotation. $\Rightarrow [(m_2 \times g) \times 1] = [(m_1 \times g) \times 3].$

Answer B!

50.   1.   Torque of F_2 about A = $F_2 \times 0 = 0.$

BECAUSE the line of action of this vertical force passes through the axis of rotation.

2.   Torque of $(m \times g)$ about an axis passing through point A =

[$(m \times g)$ VERTICAL FORCE $\times 0.2$ PERPENDICULAR (HORIZONTAL) DISTANCE TO THE AXIS OF ROTATION]
clockwise.

Note that 0.2 m is the perpendicular distance to the axis of rotation.

3.   Torque of F_1 about A = [$(F_{1,\text{ HORIZONTAL FORCE}})\times 1.2$ PERPENDICULAR (VERTICAL) DISTANCE TO THE AXIS
OF ROTATION] clockwise.

Convince yourself that the perpendicular distance is 1.2 m.

4.   Torque of F_3 about A = [$(F_{3,\text{ HORIZONTAL FORCE}})\times 0.3$ PERPENDICULAR (VERTICAL) DISTANCE TO THE AXIS
OF ROTATION] counterclockwise.

Net torque about the axis of rotation $\sum \tau = 0.$

$\Rightarrow$ [Sum of clockwise torques = Sum of counter-clockwise torques.] ABOUT THE AXIS OF ROTATION

$\Rightarrow [(m \times g) \times 0.2]$ CLOCKWISE $+ [F_1 \times 1.2]$ CLOCKWISE $= [F_3 \times 0.3]$ COUNTERCLOCKWISE.

Answer B!

52.   Torque of F_1 about A = $F_1 \times 0 = 0.$

BECAUSE the line of action of this vertical force passes through the axis of rotation. We are repeating this
again and again because students keep making the same mistake over and over again!

Answer A!

53.   1.   Torque of F_1 about A = $F_1 \times 0 = 0.$

BECAUSE the line of action of this vertical force passes through the axis of rotation. We are
repeating this again and again because students keep making the same mistake over and over again!

2.   Torque of F_2 about A =

[$F_{2,\text{ FORCE (VERTICAL)}} \times 0.2$ PERPENDICULAR (HORIZONTAL) DISTANCE TO THE AXIS OF ROTATION]
counterclockwise.

3.   Torque of $(m_2 \times g)$ about point A =

$[(m_2 \times g)_{\text{FORCE (VERTICAL)}} \times 0.6 \text{ PERPENDICULAR (HORIZONTAL) DISTANCE TO THE AXIS OF ROTATION}]$ clockwise.

4. Torque of $(m_1 \times g)$ about A =

$[(m_1 \times g)_{\text{FORCE (VERTICAL)}} \times 1.2 \text{ PERPENDICULAR (HORIZONTAL) DISTANCE TO THE AXIS OF ROTATION}]$ clockwise.

Net torque about the axis of rotation $\sum \tau = 0$.

$\Rightarrow$ [Sum of clockwise torques = Sum of counter-clockwise torques] $_{\text{ABOUT THE AXIS OF ROTATION}}$.

$\Rightarrow (m_1 \times g \times 1.2)_{\text{CLOCKWISE}} + (m_2 \times g \times 0.6)_{\text{CLOCKWISE}} = (F_2 \times 0.2)_{\text{COUNTERCLOCKWISE}}$.

Answer B!

55. The vector sum of the horizontal components, $\sum F_x = 0$. That is, the sum of the magnitudes of the horizontal components to the right = sum of the magnitudes of the horizontal components to the left.

$\Rightarrow T_1 = T_2 \times \sin 30$. Remember that the angle 30 is given with respect to the vertical.

Answer C!

56. 1. Torque of N about A = $N \times 0 = 0$.

The line of action of this vertical force passes through the axis of rotation.

2. Torque of $(m \times g)$ about A = 0 as the perpendicular distance to the axis of rotation is zero.

The line of action of this vertical force passes through the axis of rotation.

3. Torque of T_1 about A =

$[(T_1)_{\text{FORCE (HORIZONTAL)}} \times (7 \times \cos 30)_{\text{PERPENDICULAR (VERTICAL) DISTANCE TO THE AXIS OF ROTATION}}]$ counterclockwise.

Note that $7 \times \sin 30$ is parallel to tension T_1.

4. Torque of $(T_2 \times \cos 30)$ about A = 0 as the perpendicular distance to the axis of rotation is zero. The line of action of this vertical force passes through the axis of rotation.

5. Torque of $(T_2 \times \sin 30)$ about A =

$[(T_2 \times \sin 30)_{\text{FORCE (HORIZONTAL)}} \times (7 \times \cos 30)_{\text{PERPENDICULAR (VERTICAL) DISTANCE TO THE AXIS OF ROTATION}}]$ clockwise.

Net torque about the axis of rotation $\sum \tau = 0$.

$\Rightarrow$ [Sum of clockwise torques = Sum of counter-clockwise torques] $_{\text{ABOUT THE AXIS OF ROTATION}}$.

$\Rightarrow [(T_2 \times \sin 30) \times (7 \cos 30)]_{\text{CLOCKWISE}} = [T_1 \times (7 \cos 30)]_{\text{COUNTERCLOCKWISE}}$.

Answer C!

58. The vector sum of the horizontal components, $\sum F_x = 0$.

That is, the sum of the magnitudes of the horizontal components to the right = sum of the magnitudes of the horizontal components to the left.

$\Rightarrow T \times \cos 30 = F_p$.

Answer A!

59.   The vector sum of the vertical components, $\sum F_y = 0$.

That is, the sum of the magnitudes of the vertical components acting vertically up = sum of the magnitudes of the vertical components acting vertically down.

$\Rightarrow \ (T \times sin\ 30) + F_V = (m_1 \times g) + (m_2 \times g)$.

Answer B!

61.   1.   Torque of $T_1 = [T_{1,\ \text{FORCE (HORIZONTAL)}} \times R_{\ \text{PERPENDICULAR (VERTICAL) DISTANCE TO THE AXIS OF ROTATION}}$] clockwise.

2.   Torque of $T_2 =$

$[T_{2,\ \text{FORCE (VERTICAL)}} \times R_{\ \text{PERPENDICULAR (HORIZONTAL) DISTANCE TO THE AXIS OF ROTATION}}$] counterclockwise.

Net torque about the axis of rotation $\sum \tau = 0$.

$\Rightarrow$ [Sum of clockwise torques = Sum of counter-clockwise torques] $_{\text{ABOUT THE AXIS OF ROTATION}}$.

$\Rightarrow T_1 \times R = T_2 \times R$.

Answer A!

62.   1.   Torque of tension $T_1 =$

$[(T_1)_{\text{FORCE (HORIZONTAL)}} \times R_{\ \text{PERPENDICULAR (VERTICAL) DISTANCE TO THE AXIS OF ROTATION}}]$ clockwise.

2.   Torque of tension $T_2 =$

$[(T_2)_{\text{FORCE (VERTICAL)}} \times R_{\ \text{PERPENDICULAR (HORIZONTAL) DISTANCE TO THE AXIS OF ROTATION}}$] counterclockwise.

If net torque is clockwise, T_1 $\rangle_{\text{GREATER THAN}}\ T_2$.

This is so because distance at which these forces act is the same.

Do not assume this to be the general case!

Answer A!

64.   This problem is different.

Here we have a tangential force. The distance perpendicular to a tangential force is the radius!

1.   Torque of $T_1 = [(T_1)_{\text{FORCE (HORIZONTAL)}} \times R_{1\ \text{PERPENDICULAR (VERTICAL) DISTANCE TO THE AXIS OF ROTATION}}$] clockwise.

2.   Torque of $T_2 = [(T_2)_{\text{FORCE (TANGENTIAL)}} \times R_{2\ \text{PERPENDICULAR (RADIAL) DISTANCE TO THE AXIS OF ROTATION}}$] clockwise.

Both torques are clockwise.

$\Rightarrow$ Net torque is clockwise.

Answer A!

65.   1.   Torque of normal force = $N \times 0 = 0$.
The line of action of this vertical force passes through the axis of rotation.

2.   Torque of $(m \times g) = 0$.
The line of action of this vertical force passes through the axis of rotation.

3.   Torque of force, $F = [F_{\text{FORCE (HORIZONTAL)}} \times R_{\text{RADIUS OF THE BALL (PERPENDICULAR DISTANCE)}}]$.
Force F will rotate the basketball in the counterclockwise direction (if viewed from the top).

Answer B!

67.   Since weight is now off the axis of rotation, it will contribute to an unbalanced torque that will tip the ball.
Answer A!

68.   1.   Torque of the weight of the cylindrical mass $(m \times g)$ about an axis passing through its center = $[(m \times g) \times 0] = 0$.
The line of action of this vertical force passes through the axis of rotation.

2.   Torque of tension = $[T_{\text{FORCE (VERTICAL)}} \times R_{\text{PERPENDICULAR (HORIZONTAL) DISTANCE TO THE AXIS OF ROTATION}}]$ clockwise.

Net torque = $T \times R$ clockwise.

(Note: As the thread unwinds, the cylinder will increase in angular speed.
Angular acceleration will be non-zero. So tension T is not equal to m_1g!)
Answer A!

70.   The vector sum of the horizontal components, $\sum F_x = 0$.
That is, the sum of the magnitudes of the horizontal components to the right = sum of the magnitudes of the horizontal components to the left.
$\Rightarrow F_2 = F_1 \times \cos \theta$.
Answer D!

71.   The vector sum of the horizontal components, $\sum F_x = 0$.
That is, the sum of the magnitudes of the horizontal components to the right = sum of the magnitudes of the horizontal components to the left.
$\Rightarrow F_{V1} + F_{V2} = T \times \cos 60$.
Answer B!

73.   The vector sum of the horizontal components, $\sum F_x = 0$. That is, the sum of the magnitudes of the horizontal components to the right = sum of the magnitudes of the horizontal components to the left.
$\Rightarrow T_1 \times \cos \theta_1 = T_2 \times \cos \theta_2$.
Answer A!

74.   1.   Torque of $(T_1 \times cos\ \theta_1)$ about A = 0.

The line of action of this horizontal force passes through the axis of rotation.

2.   Torque of $(T_1 \times sin\ \theta_1)$ about A = 0.

The line of action of this vertical force passes through the axis of rotation.

3.   Torque of $(m_1 \times g)$ about A =

$[(m_1 \times g)$ FORCE (VERTICAL) $\times\ 0.2$ PERPENDICULAR (HORIZONTAL) DISTANCE TO THE AXIS OF ROTATION$]$ clockwise.

4.   Torque of $(m_2 \times g)$ about A =

$[(m_2 \times g$)FORCE (VERTICAL) $\times\ 0.5$ PERPENDICULAR (HORIZONTAL) DISTANCE TO THE AXIS OF ROTATION$]$ clockwise.

5.   Torque of $(T_2 \times cos\ \theta_2)$ about A = $[T_2 \times cos\ \theta_2 \times\ 0] = 0$.

The line of action of this horizontal force passes through the axis of rotation.

6.   Torque of $(T_2 \times sin\ \theta_2)$ about A =

$[(T_2 \times sin\ \theta_2$)FORCE (VERTICAL) $\times\ 1$ PERPENDICULAR (HORIZONTAL) DISTANCE TO THE AXIS OF ROTATION$]$ counterclockwise.

Net torque about the axis of rotation $\sum \tau = 0$.

$\Rightarrow$ [Sum of clockwise torques = Sum of counter-clockwise torques] ABOUT THE AXIS OF ROTATION.

$\Rightarrow [\ (m_1 \times g \times 0.2) + (m_2 \times g \times 0.5)]_{\text{CLOCKWISE}} = [T_2 \times sin\ \theta_2 \times 1]$ COUNTERCLOCKWISE.

Answer A!

76.   The vector sum of the horizontal components, $\sum F_x = 0$.

That is, the sum of the magnitudes of the horizontal components to the right = sum of the magnitudes of the horizontal components to the left. $\Rightarrow F_f = F_{\text{wall}}$.

Answer B!

77.   The vector sum of the vertical components, $\sum F_y = 0$.

That is, the sum of the magnitudes of the vertical components acting vertically up = sum of the magnitudes of the vertical components acting vertically down.

$\Rightarrow N = W$.

Answer C!

79.   1.   Torque of $(m \times g)$ on the right about A =

$[(m \times g$) FORCE (VERTICAL) $\times R$ PERPENDICULAR (HORIZONTAL) DISTANCE TO THE AXIS OF ROTATION $]$ clockwise.

2.   Torque of $(m \times g)$ on the left about A =

$[(m \times g$) FORCE (VERTICAL) $\times R$ PERPENDICULAR (HORIZONTAL) DISTANCE TO THE AXIS OF ROTATION $]$ counterclockwise. These torques cancel each other!

Answer A!

80.   Torque of $(m_1 \times g) =$

[$(m_1 \times g)$ FORCE (VERTICAL) $\times R$ PERPENDICULAR (HORIZONTAL) DISTANCE TO THE AXIS OF ROTATION] clockwise.
Answer A!

82.   The vector sum of the horizontal components, $\sum F_x = 0$.

That is, the sum of the magnitudes of the horizontal components to the right = sum of the magnitudes of the horizontal components to the left.

$\Rightarrow F_f = F_{\text{wall}} \times cos\ \theta$.
Answer A!

83.   The vector sum of the vertical components, $\sum F_y = 0$.

That is, the sum of the magnitudes of the vertical components acting vertically up = sum of the magnitudes of the vertical components acting vertically down.

$\Rightarrow (F_{\text{wall}} \times sin\ \theta) + N = m \times g$.
Answer C!

85.
  1.   Torque of F_V about A = $F_V \times 0 = 0$.
     The line of action of this vertical force passes through the axis of rotation.
  2.   Torque of F_H about A = 0.
     The line of action of this horizontal force passes through the axis of rotation.
  3.   Torque of $(m_2 \times g)$ about A =
     [$(m_2 \times g)$ FORCE (VERTICAL) $\times 2.5$ PERPENDICULAR (HORIZONTAL) DISTANCE TO THE AXIS OF ROTATION] clockwise.
  4.   Torque of $(m_1 \times g)$ about A =
     [$(m_1 \times g)$ FORCE (VERTICAL) $\times 5$ PERPENDICULAR (HORIZONTAL) DISTANCE TO THE AXIS OF ROTATION] clockwise.
  5.   Torque of $(T \times cos\ 30)$ about A = [$(T \times cos\ 30) \times 0$] = 0.
     The line of action of this horizontal force passes through the axis of rotation.
  6.   Torque of $(T \times sin\ 30)$ about A =
     [$(T \times sin\ 30)$ FORCE (VERTICAL) $\times 5$ PERPENDICULAR (HORIZONTAL) DISTANCE TO THE AXIS OF ROTATION] counterclockwise.

Net torque about the axis of rotation $\sum \tau = 0$.

$\Rightarrow$ [Sum of clockwise torques = Sum of counter-clockwise torques] ABOUT THE AXIS OF ROTATION. ·

$\Rightarrow [(m_2 \times g \times 2.5) + (m_1 \times g \times 5)]$ CLOCKWISE = [$T \times sin\ 30 \times 5$] COUNTER-CLOCKWISE.
Answer C!

86.   The vector sum of the horizontal components, $\sum F_x = 0$.

That is, the sum of the magnitudes of the horizontal components to the right = sum of the magnitudes of the horizontal components to the left.

$\Rightarrow F_H = T \times cos\ 30$.
Answer B!

1. Any object that moves in a circle has a net force acting on it. This is because the direction of the velocity vector is changing. If there were no force acting, the object would continue with the same velocity: Newton says so! Note that the magnitude of the velocity does not change in this problem.
    Answer B!

2. The direction of angular acceleration is perpendicular to the plane of the circle.
    Notes about angular acceleration:
    Angular acceleration is *different* from linear acceleration. Angular acceleration gives you a factor by which the angular speed changes.
    If angular acceleration is 3 rad/s^2, then the angular speed is increasing by 3 rad/s every second. If it started from rest, it will rotate at 3 rad/s after one second. After 2 s, it will be rotating at an angular speed of 6 rad/s if the angular acceleration is a constant. It will rotate faster and faster!

    Angular acceleration:
    α ANGULAR ACCELERATION $= [(\omega_f - \omega_0)$CHANGE IN ANGULAR VELOCITY $/ t$ TIME $]$.
    If angular velocity is a constant, angular acceleration will be zero.
    Answer C!

4. Radial acceleration or centripetal acceleration is:
    a CENTRIPETAL ACCELERATION $= [v$LINEAR SPEED$^2 / R$ RADIUS $]$.
    Note that v LINEAR SPEED $= R$ RADIUS $\times \omega$ ANGULAR SPEED. This is another relation that you can never forget!
    $\Rightarrow a = [v^2 / R] = [(R \times \omega_f)^2 / R] = [R^2 \times \omega_f^2] / R =$
    $\quad\quad R \times \omega_f^2 = R \times [\omega_0 + (\alpha \times t)]^2$.
    Radial acceleration changes with time only if angular acceleration α is non-zero.
    Answer B!

5. ω_f FINAL ANGULAR SPEED $\quad\quad\quad = [30$ rev/sec$] \times [2\pi$ rad/rev$] = [60\pi$ rad/sec$]$.
    ω_0 INITIAL ANGULAR SPEED $\quad\quad\quad = [20$ rev/sec$] \times [2\pi$ rad/rev$] = [40\pi$ rad/sec$]$.
    The change from a "low" initial value to a higher final value occurs within a time of 2 s. There is a *change* of angular speed. That is, the system is accelerating.
    There is a non-zero angular acceleration α.

    $t = 2$ sec.
    By definition, angular acceleration:
    α ANGULAR ACCELERATION $= [(\omega_f - \omega_0)$ CHANGE IN ANGULAR SPEED $/ t$ TIME $]$
    $\quad\quad\quad = [(60\pi - 40\pi) / 2] = 10\pi$ rad/s$^2 = 10 \times 3.14 = 31.4$ rad/s^2.
    That is, the system increases the rotational speed by 31.4 rad/s every second!
    Another important thing that you should not forget is that with the increasing rotational speed the object turns through more angle in the same unit time.
    Answer A!

7. Please look at the solution to problem **5**.

a_T, TANGENTIAL ACCELERATION $= [R$ RADIUS $\times \alpha$ ANGULAR ACCELERATION$]$.

That is, the tangential acceleration really depends upon the distance R! For particles at different radii, tangential acceleration is different! Note that angular acceleration α is the same for all the particles of a rigid body. Angular velocity is also the same for all particles of a rigid body. Linear speed v is *not* the same for all the particles of a rigid body! It depends on how far from the axis they are located.

Tangential acceleration, $a_T = [R \times \alpha]$. $\Rightarrow a_T = [0.2 \text{ m} \times 10\pi \text{ rad/s}^2] = 2\pi \text{ m/s}^2$.

Note: Radians are pure numbers without any dimensions. We worry about mass (M), length (L), and time (T). Since radians do not fall into it [MLT], we generally do not include it.
Answer C!

8. Look for ω_f.

$\Rightarrow \omega_f$ FINAL ANGULAR VELOCITY $= \omega_0$ INITIAL ANGULAR VELOCITY $+ (\alpha$ ANGULAR ACCELERATION $\times t$ TIME$)$.
From solution 5, $\omega_0 = 40\pi \text{ rad/sec}$, $\alpha = 10\pi \text{ rad/s}^2$, and $t = 1.5$ s.
$\Rightarrow \omega_f = (40\pi) + (10\pi \times 1.5) = 55 \pi \text{ rad/s}$.
Answer A!

10. The question says that it makes 50 revolutions in 5 seconds. It *does not* mean that its final angular speed is [50 rev / 5 s]!
Instead, what we are told is that: angle $\theta = [50 \text{ revolutions}] \times [(2\pi \text{ rad}) / (1 \text{ rev})]$
$= 100\pi$ radians. This much angle is turned in 5 s. $\Rightarrow t = 5$ s.
Use the relation:

$\Delta\theta$ ANGLE $= (\omega_0$ INITIAL ANGULAR VELOCITY $\times t$ TIME$) + (\frac{1}{2} \times \alpha$ ANGULAR ACCELERATION $\times t^2)$.
The initial angular speed is zero. $\Rightarrow \Delta\theta = 0 + (\frac{1}{2} \times \alpha \times t^2)$.
$\Rightarrow \alpha = (2 \times \Delta\theta) / t^2 = [(2 \times 100\pi)] / 5^2 = 8\pi \text{ rad/s}^2$.
Answer A!

11. $\theta = [50 \text{ revolutions}] \times [(2\pi \text{ rad}) / (1 \text{ rev})] = 100\pi \text{ rad}$. This much angle is turned in 5 s. $\Rightarrow t = 5$ s. Use the relation:

$\Delta\theta$ ANGLE $= (\omega_0$ INITIAL ANGULAR VELOCITY $\times t$ TIME$) + (\frac{1}{2} \times \alpha$ ANGULAR ACCELERATION $\times t^2)$.
The initial angular speed is zero. $\Delta\theta = 0 + (\frac{1}{2} \times \alpha \times t^2)$.
$\Rightarrow \alpha = (2 \times \Delta\theta) / t^2 = [(2 \times 100\pi)] / 5^2 = 8\pi \text{ rad/s}^2$.
Now are ready to use: $\omega_f = \omega_0 + (\alpha \times t)$. $\Rightarrow \omega_f = 0 + (8 \pi \text{ rad/s}^2 \times 3 \text{ s}) = 24 \pi \text{ rad/s}$.
Answer B!

13. In the first five seconds, the object is accelerating. In the next 2 s, it is going at constant angular speed. The number of rotations made in 7 seconds equals the number of rotations it made in the first 5 seconds + the number of rotations it made in the next 2 seconds.

Total revolutions = 50 revolutions + # revolutions in the next 2 seconds.

Note that 50 revolutions was given to us in the problem. Beyond $t = 5$ s, it is going at a constant angular speed. Let us find that angular speed at the end of 5 s.

Use: $\omega_f = \omega_0 + (\alpha \times t)$. $\Rightarrow \omega_f = 0 + (8\pi \times 5 \text{ s}) = 40\pi \text{ rad/s}$.

The number of rotations it makes in the next 2 seconds when the object is rotating at a constant angular speed of $40\pi \text{ rad/s}$ is $\Delta\theta = \omega \times t = [40 \ \pi \text{ rad/s}] \times [2 \text{ s}] = 80\pi \text{ radians}$.

The number of revolutions in 80π radians $= (80\pi) / (2\pi) = 40$ revolutions.

So the total number of revolutions in the first seven seconds $= 50 + 40 = 90$ revolutions.

Answer A!

14. Radial acceleration is $a_R = a = [v^2 / R] = [(R \times \omega_f)^2 / R] = [R^2 \times \omega_f^2] / R = R \times \omega_f^2$
$= R \times [\omega_0 + (\alpha \times t)]^2$.
$\Rightarrow a_R = R \times [0 + (\alpha \times t)]^2$. $\Rightarrow a_R = R \times \alpha^2 \times t^2$.
Answer A!

16. r_1 is the distance in m from the axis of rotation O to the position of mass m_1 at P.
r_2 is the distance in m from the axis of rotation O to the position of mass m_2 at Q.
Answer A!

17. Answer B!

19. Answer A!

20. Answer A!

22. $0 = x_{cm} = \{[m_1 \times (-x)] + [m_2 \times (l - x)]\} / (m_1 + m_2)$.
Answer A!

23. $0 = (-m_1 \times x) + (m_2 \times l) - (m_2 \times x)$.
[Solve for $x = (m_2 \times l) / (m_1 + m_2)$.]
Answer A!

25. If $x = [m_2 / (m_1 + m_2)] \times l$, then
$l - x = l - [m_2 / (m_1 + m_2)] \times l = l \times \{1 - [m_2 / (m_1 + m_2)]\} = l \times [(m_1 + m_2 - m_2) / (m1 + m_2)]$
$\Rightarrow l - x = l \times [m_1 / (m_1 + m_2)]$.
Answer A!

26. Now we have that $x = (m_2 \times l) / (m_1 + m_2)$ and that $l - x = (m_1 \times l) / (m_1 + m_2)$.
Answer A!

28.   The rotational inertia about point $P = (m_1 \times 0^2) + (m_2 \times l^2) = (m_2 \times l^2)$.
       Answer B!

29.   The rotational inertia about an axis passing through P and Q $= (m_1 \times 0^2) + (m_2 \times 0^2) = 0$.
       The rotational inertia along an axis passing through PQ is zero because the masses are placed at the axis so
       they do not swing at all during the rotation and have no inertial effect. The assumption we make is that they
       are point masses.
       Answer C!

31.   Using the parallel axis theorem,
       $I_P = I_{CM} + (m \times r^2)$,
       where moment of inertia about the center of mass $I_{CM} = m \times R^2$ for a ring if R is the radius.
       In this case, $I_{CM} = 2 \text{ kg} \times (0.1 \text{ m})^2 = 0.02 \text{ kg} \times \text{m}^2$.
       The distance to the center of the ring from point $P = 0.5 + 0.1 = 0.6$ m.
       $m \times r^2 = 2 \text{ kg} \times (0.5 \text{ m} + 0.1 \text{ m})^2$.
       Note again that *the distance to the axis of rotation = the length of the wire + the radius of the ring*.
       $I_P = I_{CM} + (m \times r^2)$ becomes $I_{cm} = 2 \times (0.1)^2 + 2 \times (0.5 + 0.1)^2 = 0.74 \text{ kg} \times \text{m}^2$.
       Answer B!

32.   Center of mass of a two-body system is an important concept for you to understand. There is a lot of math
       involved. Let us try to understand it better!
       Common sense: Note that the chlorine mass (m_2) is bigger than the hydrogen mass (m_1). That is, the center
       of mass of the two masses is not at the center! The center of mass of HCl is closer to chlorine than to
       hydrogen!

             STEP 1:   Let d be the distance between the two masses (m_1) and (m_2).
                            Assume the origin is chosen at the center of mass.
             STEP 2:   Distance of the center of mass from the hydrogen mass
                            $= (m_2 \times d) / (m_1 + m_2) = r_1$.
             STEP 3:   Distance of the center of mass from the chlorine atom
                            $= (m_1 \times d) / (m_1 + m_2) = r_2$.

       THINK THROUGH! Think about what these equations (steps 2 and 3) mean. We know that the
       center of mass is between the masses, and that it is closer to the center of the larger mass. These
       equations tell us, for example, that the ratio of r_1 to the total distance $d (= r_1 / d)$ is equal to the total
       mass fraction of $m_2 = m_2 / (m_1 + m_2)$.
       The moment of inertia of the system about the axis perpendicular to the line joining the two masses through
       the center of mass of H and Cl is:
             STEP 4: $I_{CM} = (m_1 \times r_1^2) + (m_2 \times r_2^2)$.
       Now use all the math you have learned to go through the following.
       $I_{CM} = (m_1 \times r_1^2) + (m_2 \times r_2^2) = \{m_1 \times [m_2 \times d / (m_1 + m_2)]^2\} + \{m_2 \times [m_1 \times d / (m_1 + m_2)]^2\}$.

       $\Rightarrow I_{CM} \qquad = [(m_1 \times m_2^2 \times d^2) / (m_1 + m_2)^2] + [(m_2 \times m_1^2 \times d^2) / (m_1 + m_2)^2]$.

The denominator is the same. We add the numerators.
$$= [(m_1 \times m_2{}^2 \times d^2) + (m_2 \times m_1{}^2 \times d^2)] / (m_1 + m_2)^2.$$
Now take the common factor $m_1 \times m_2 \times d^2$ outside.
$$\Rightarrow I_{CM} = m_1 \times m_2 \times d^2 \,[\,(m_2 + m_1) / (m_1 + m_2)^2\,].$$
Now cancel $(m_1 + m_2)$ once.
$$\Rightarrow I_{CM} = (m_1 \times m_2 \times d^2) / (m_1 + m_2).$$
Rearranging we get: $I_{CM} = [(m_1 \times m_2) / (m_1 + m_2)] \times d^2.$
This is just the answer to problem **27**!
m_1 = mass of hydrogen = 1.7×10^{-27} kg, $m_2 = 35 \times 1.7 \times 10^{-27}$ kg. Note again that chlorine is 35 times heavier than hydrogen. $d = 1.2 \times 10^{-11}$ m.
Substitute the known quantities and solve.
$$\Rightarrow I_{CM} = 24 \times 10^{-50}\ \text{kg} \times \text{m}^2.$$
Answer A!

34. Moment of inertia of the solid cylinder about an axis (along the axis of the cylinder) passing through its center $I_{\text{CYLINDER}} = m_{\text{MASS}} \times R_{\text{RADIUS}}{}^2 / 2.$
Note that a cylinder is nothing but a FAT disc!
The moment of inertia has nothing to do with the length of the cylinder.
$$\Rightarrow I = 10\ \text{kg} \times (0.05\ \text{m})^2 / 2 = 1.25 \times 10^{-2}\ \text{kg} \times \text{m}^2.$$
Answer A!

35. The symmetry is broken by the unequal masses of 100 grams, 200 grams, 300 grams, and 400 grams at the corners of the square. Because of this, the center of the square will not coincide with the center of mass.
If you apply your common sense, we immediately find that the position of the center of mass has $x_{CM} = 0$ and y_{CM} is not equal to zero. [Note that: $100 + 400 = 500$, and $200 + 300 = 500$! That is why!]

If you apply the formula: $x_{CM} = [(m_1x_1) + (m_2x_2) + (m_3x_3) + (m_4x_4)] / (m_1 + m_2 + m_3 + m_4)$
$= \{[100 \times (-10)] + [200 \times 10] + [300 \times 10] + [400 \times (-10)] / (100 + 200 + 300 + 400)\} = 0!$

If you do not believe that y_{CM} is *not* equal to zero, try calculating y_{CM} by the same recipe.
Answer B!

37. The moment of inertia of the system about an axis passing through the center O:
$I_{\text{ROTATIONAL INERTIA}} = (m_1 \times r_1{}^2) + (m_2 \times r_2{}^2) + (m_3 \times r3^2) + (m_4 \times r_4{}^2).$ Note that O is at the center of the square so the distance to all of the masses is the same! That is $r_1 = r_2 = r_3 = r_4 = r = 14.14$ cm.
$$\Rightarrow I_{\text{ROTATIONAL INERTIA}} =$$
$$[100\ \text{g} \times (14.14\ \text{cm})^2] + [200\ \text{g} \times (14.14\ \text{cm})^2] + [300\ \text{g} \times (14.14\ \text{cm})^2] + [400\ \text{g} \times (14.14\ \text{cm})^2\,].$$
$I_O = (100 + 200 + 300 + 400)\ \text{grams} \times (14.14\ \text{cm})^2$
$= 200{,}000\ \text{g} \times \text{cm}^2 = 0.020\ \text{kg} \times \text{m}^2$, since 1 kg = 1,000 g and 1 m = 100 cm.
Note that $(1\ \text{m})^2 = 100\ \text{cm} \times 100\ \text{cm} = 10{,}000\ \text{cm}^2.$
Answer D!

38.   The moment of inertia of the rod about one end $I_{\text{ROTATIONAL INERTIA}} = [m_{\text{MASS}} \times l_{\text{LENGTH}}^2] / 3$
       $= [(0.6 \text{ kg}) \times (1.0 \text{ m})^2] / 3 = 0.2 \text{ kg} \times \text{m}^2$.
       Do not forget that moment of inertia depends upon the axis of choice!
       Answer C!

40.   The moment of inertia of the system is $= (I_{\text{ROD}}) + (I_{\text{RING}})$ about the nail at point A.
       STEP 1:   Find I_{ROD} about the nail.
                     The moment of inertia of the rod about one end $I_{\text{ROTATIONAL INERTIA}}$
                     $= [m_{\text{MASS}} \times l_{\text{LENGTH}}^2] / 3 = [(0.6 \text{ kg}) \times (1.0 \text{ m})^2] / 3 = 0.2 \text{ kg} \times \text{m}^2$. [From problem **38**].
       STEP 2:   Find I_{RING} about the nail.
                     The moment of inertia of the ring about the nail
                     $= [I_{\text{CM}} \text{ of the ring}] + [m \times (1.0 + 0.1)_{\text{DISTANCE TO THE AXIS OF ROTATION}}^2]$.
                     $\Rightarrow I_{\text{RING}} = [(2 \text{ kg}) \times (0.1 \text{ m})^2] + [(2 \text{ kg}) \times (1.0 \text{ m} + 0.1 \text{ m})^2] = 2.44 \text{ kg} \times \text{m}^2$.
       STEP 3:   Now we are ready to find the inertia of the system (ring and the rod) about point A.
                     $I_{\text{SYSTEM}} = 0.2 + 2.44$. That is, $I_{\text{SYSTEM}} = 2.64 \text{ kg} \times \text{m}^2$.
       Answer C!

41. $F \times cos\ \theta$ is along the horizontal. That force will not produce a torque.
       Torque about point O $= (F \times sin\ \theta) \times r = F \times r \times sin\ \theta$.
       Answer A!

43.   Translation is produced by forces. Rather an Archimedean concept!
       Net force produces translational acceleration!
       Answer A!

44.   Torque produces rotation. Net torque produces rotational acceleration.
       Answer B!

46.   Answer B!

47.   Answer B!

49.   Answer B!

50.   Answer A!

52.   For an object to be in equilibrium, it should not have a net force acting on it nor a net torque. A net force
       will produce translational acceleration. A net torque will produce rotational acceleration.
       An interesting fact: If there is zero net force on a body, and the net torque about one point is zero, it is zero
       about all points automatically! So actually Answer C is more than necessary and sufficient.
       Answer C!

53. Net force along $y = 0$. $\Rightarrow (W_1 + W + W_2)_{\text{VERTICALLY DOWN}} = (R_1 + R_2)_{\text{VERTICALLY UP}}$.

Net torque is zero.

$\Rightarrow [(W_1 \times \text{PE}) + (W \times \text{PO}) + (W_2 \times \text{PF})]_{\text{CLOCKWISE}} = [(R_1 \times \text{PL}) + (R_2 \times \text{PM})]_{\text{COUNTERCLOCKWISE}}$.

Answer A!

54. It is very hard to find a recipe for problems such as this. The important things to remember are that:

1. The ladder has a tendency to slide to the left, so force of friction F_f is to the right.
2. N_1 is the normal force exerted at the base of the ladder. $\Rightarrow F_f = \mu \times (\text{Normal force}) = \mu N_1$.
3. The weight of the ladder acts halfway at a distance of 2.5 meters from the base along the ladder.
4. The weight of the person acts at a distance of x meters from the base.

If the ladder does not rotate (flip) or slide, we expect the following to be true for the body in question, the ladder: The two parts that the students should get without too much trouble are:

STEP 1: $\sum F_x = 0$.

$\Rightarrow [F_{\text{WL}}]_{\text{LEFT}} = [\mu \times N_1]_{\text{RIGHT}} = F_f$ **(1)**

STEP 2: $\sum F_y = 0$.

$\Rightarrow [60g + 20g]_{\text{VERTICALLY DOWN}} = [N_1]_{\text{VERTICALLY UP}}$ or $N_1 = 80g = 80 \times 10 = 800$ N.   **(2)**

Note that the sign of g is not negative because we already took the direction of the weight into account!

Once we know the normal force N_1 [from **(2)**] and the coefficient of friction μ, we know the force of friction.

$\Rightarrow F_f = \mu \times N_1 = 0.4 \times 800 = 320$ N.

By equation **(1)**, $F_f = F_{\text{WL}} = 320$ N.

$\Rightarrow$ Force of wall on the ladder is $F_{\text{WL}} = 320$ N as $\sum F_x = 0$.

STEP 3: Here comes the difficult part. Pay close attention.

Let us determine the torque of each of the forces involved about the axis of rotation about the base of the ladder, point A.

1. Torque of $N_1 = N_1 \times 0 = 0$.
2. Torque of $\mu \times N_1 = \mu \times N_1 \times 0 = 0$.
3. Torque of the weight of the person

= $W_{\text{P}} \times$ perpendicular distance to point A

= $60g \times x \cos \theta$ (clockwise about point A).

Note that hypotenuse is x and the angle with respect to the horizontal is θ.

Weight is acting vertically down. This makes the horizontal distance to the axis of rotation = $x \times \cos \theta$.

4. Torque of the weight of the ladder about point A

= $W_{\text{L}} \times$ perpendicular distance to point A

= $20g \times 2.5 \cos \theta$ (also clockwise).

Note that the weight of a uniform ladder acts exactly in the middle.

Weight acts vertically down. This makes the horizontal distance to the axis of rotation

= $2.5 \times \cos \theta$.

5. Torque of the force F_{WL} about A = $F_{\text{WL}} \times \text{BO} = F_{\text{WL}} \times 5 \times \sin \theta$ (counterclockwise). Note

that the force F_{WL} is horizontal so the vertical distance to the axis of rotation
= $5 \times \sin \theta$. Hypotenuse is 5 m.

Set the sum of the clockwise torques = sum of the counterclockwise torques.

$\Rightarrow \{[(60g) \times (x \times \cos \theta)] + [(20g) \times (2.5 \times \cos \theta)]\}_{CLOCKWISE}$

$= [(F_{WL}) \times (5 \times \sin \theta)]_{COUNTER\text{-}CLOCKWISE}$.

To find θ

$\cos \theta = 3 / 5 = 0.6$.

$\Rightarrow \theta = 53.1$ degrees.

Substitute the known quantities.

$(60 \times 10 \times x \times \cos 53) + (20 \times 10 \times 2.5 \times \cos 53) = (320 \times 5 \times \sin 53)$.

$\Rightarrow (360 \times x) + 300 = 1280$.

$\Rightarrow x = (1280 - 300) / 360 = 2.72$ m.

Answer D!

55. Fast Solution: At the situation where the meter stick is barely lifted off the support at point C, there is no
contact at C. The balance of torques is: $\sum \tau = 0$ about point D gives us that

$[W_M \times (0.1)]_{COUNTERCLOCKWISE} = [W_B \times (0.4)]_{CLOCKWISE}$.

$\Rightarrow W_B = [W_M \times (0.1)] / (0.4)$

$= [2 \text{ N} \times (0.1)] / (0.4)$

$= 0.5$ N.

All the details you (do not?) want:

Note again that we are given a meter stick. That is the length is 1 m. The supports are placed symmetrically
about the center.

$CD = 0.2$ m = $\{(1 \text{ m}) - [2 \times (AC)]\} = \{(1 \text{ m}) - [2 \times (BD)]\}$.

$\Rightarrow AC = BD = 0.4$ m.

Trick: When the meter stick is ready to flip over, normal force exerted at point C will be zero. At that
instant, $\sum F_x = 0$, $\sum F_y = 0$, and $\sum \tau = 0$.

$\sum F_x = 0$ gives us nothing.

$\sum F_y = 0$ gives us that the normal force at point D = $N_D = W_M + W_B$, where W_M is the weight
of the meter stick and W_B is the weight hanging at point B.

$\Rightarrow [N_D]_{VERTICALLY\ UP} = [2 \text{ N} + W_B]_{VERTICALLY\ DOWN}$.

$\sum \tau = 0$ about point D gives us that

$[W_M \times (0.1)]_{COUNTERCLOCKWISE} = [W_B \times (0.4)]_{CLOCKWISE}$.

$\Rightarrow W_B = [W_M \times (0.1)] / (0.4) = [2 \times (0.1)] / (0.4) = 0.5$ N.

Answer C!

56. $F_1 \times d_1$ is clockwise and $F_2 \times d_2$ is counterclockwise.
Answer A!

58.   The non-uniform stick balances when the support is at point G.
⇒ The center of gravity of the stick acts at point G. AG = 0.6 m.
When the support is placed at point C, the center of gravity of the stick is to the right of the support. This will produce a torque in the clockwise direction. Do not forget that $0.2g = 0.2 \text{ kg} \times 10 \text{ m/s}^2 = 2$ N force.
Balance the torques.
⇒ $[(0.2g) \times (AC)]_{\text{COUNTERCLOCKWISE}} = [W_{\text{STICK}} \times CG]_{\text{CLOCKWISE}}.$
Let us look at what is on the right side.
⇒ $[W_{\text{STICK}} \times CG] = W_{\text{STICK}} \times (AG - AC) = W_{\text{STICK}} (0.6 - 0.5).$
⇒ $W_{\text{STICK}} \times (0.1).$
Now go back.
⇒ $(2 \text{ N}) \times (0.5) = W_{\text{STICK}} \times (0.1).$
⇒ $W_{\text{STICK}} = 1.0 / 0.1 = 10$ N. Same as 1 kg.
Answer A!

59.   Net torque about point C: $(F \times AC) + (F \times BC) = F \times (AC + BC) = F \times (AB)$ counterclockwise.
Net torque about point D: $F \times AD$ is clockwise. $F \times BD$ is counterclockwise.
Net torque about point D $= -[F \times (AD)] + [F \times (BD)] = F \times [BD - AD] = F \times AB$ counterclockwise.
Net torque about point E: $F \times (EB)$ is clockwise. $F \times (AE)$ is counterclockwise.
Net torque about point E $= [F \times AE] + [- (F \times EB)] = F \times [AE - EB] = F \times AB$ counterclockwise.
Equal and opposite forces applied to a body, with a separation d between their lines of action, are called a "force couple," and the net torque they produce is the same, $\tau_{\text{NET TORQUE}} = Fd$, no matter what center of rotation is chosen for the calculation.
Answer B!

61.   If two forces are equal in magnitude but opposite in direction, the vector sum of the forces (called the net force) = 0. That means linear acceleration will be zero. There is, however, a net torque for this problem. Hence there is a rotating motion.
Answer C!

62.   Center of gravity and center of mass coincide for a body at the surface of the earth, where gravity is uniform.
Answer A!

64.   Angular momentum = $I \times \omega$, along the axis.
Note that in rotational motion the "damage" the object can cause depends upon both how fast it is rotating and how its mass is distributed around the axis of rotation. Also note that this equation is easy to remember: In translational motion, linear momentum $p = m_{\text{MASS}} v_{\text{VELOCITY}}$; in rotational motion, angular momentum $L = I_{\text{MOMENT OF INERTIA}} \times \omega_{\text{ANGULAR VELOCITY}}.$
Answer B!

65. Angular momentum $= L = I \times \omega$. For a particle at a distance R, $I = m \times R^2$.

$\Rightarrow L = I \times \omega = (m \times R^2) \times \omega = m \times R^2 \times (v / R)$

$\Rightarrow L = m \times R \times v = R \times p$.

Remember that $m \times v$ is the magnitude of linear momentum.

Answer C!

67. Now the small disk is rotating about its center of mass.

Thus, $I_{\text{TOTAL}} = I_{\text{SMALL DISC}} + I_{\text{BIG DISC}}$.

STEP 1: Radius of the small disc $= R / 2$.

Its inertia is then $I_{\text{SMALL}} = [m \times (R / 2)_{\text{RADIUS of SMALL DISC}}^2] / 2$.

$\Rightarrow I_{\text{TOTAL}} = [m \times (R / 2)^2] / 2 + I_{\text{BIG DISC}}$.

$\Rightarrow I_{\text{TOTAL}} = [(m \times R^2) / 8] + I_{\text{BIG DISC}}$.

STEP 2: To find $I_{\text{BIG DISC}}$: $I_{\text{BIG DSIC}} = I_{\text{CM}} + Md^2$.

The distance of the center of the big disc from the axis of rotation $d = R / 2$.

$I_{\text{BIG}} = [(M \times R^2) / 2] + [M \times (R / 2)^2] = [(M \times R^2) / 2] + [(M \times R^2) / 4] = (3 \times M \times R^2) / 4$.

$I_{\text{TOTAL}} = [(m \times R^2) / 8] + [(3/4) \times M \times R^2]$.

This is the same as: $[(3/4) \times M \times R^2] + [(m \times R^2) / 8]$.

$I_{\text{TOTAL}} = \{[(3/4) \times M] + [m / 8]\} \times R^2$.

Answer A!

68. Fast Track Solution:

$I_{\text{SYSTEM}} = I_{\text{BIG DISC}} - I_{\text{REMOVED DISC}}$.

$I = [(1/2)mR^2]_{\text{BIG DISC}} - [(1/2)(m / 4)(R / 2)^2]_{\text{REMOVED MASS INERTIA}} - [(m / 4)(R / 2)^2]_{\text{DIST. TO THE AXIS}}$

OF ROTATION

WITH DETAILS:

When area is removed, the mass is less. How do we know the mass of the area removed?

Area density $= m / \text{total area} = m / (\pi \times R^2)$.

Area of the portion removed $= \pi \times \text{radius}^2$.

Area of the portion removed $= [\pi \times (R / 2)^2] = [(\pi \times R^2) / 4]$.

The mass of this much area that is removed is then: Area density $\times$ area removed $=$

$[m / (\pi \times R^2)] \times [(\pi \times R^2) / 4] = m / 4$.

$\Rightarrow$ Mass remaining $=$ Initial mass $-$ mass removed $= m - (m / 4) = (3/4) \times m$.

Moment of inertia of this mass that is removed (note that the mass removed is off-center):

$I_{\text{REMOVED MASS}} = [(m / 4) \times (R / 2)^2 / 2] + [(m / 4) \times (R / 2)^2] = (3/32) \times m \times R^2$.

$I_{\text{SYSTEM}}\quad = I_{\text{BIG DISC}} - I_{\text{REMOVED MASS}}$

$\quad\quad\quad\quad = [(m \times R^2) / 2] - [(3/32) \times m \times R^2] = [(13/32) \times m \times R^2]$.

Answer A!

70.   The mass m_2 ($= 4\,m_1$) is at rest instantaneously BUT ACCELERATING!

Net force can be non-zero even for a body having $v = 0$ m/s.

$\Rightarrow$ $[m_2 \times g]_{\text{BIGGER FORCE VERTICALLY DOWN}} - [T]_{\text{SMALLER FORCE VERTICALLY UP}} = [(4 \times m_1) \times a]_{\text{NET FORCE}}.$

Note that there also is a net torque $= T \times R$, which gives the cylinder an angular acceleration!

Answer C!

71.   When the mass m_2 ($= 4 \times m_1$) is released from rest, the gravitational pull makes the mass m_2 accelerate down.

Then the sum of all the forces acting down will be greater than the sum of all the forces acting up.

STEP 1: $\sum F_y = m \times a \Rightarrow [(m_2 \times g) - T] = m_2 \times a.$

$\Rightarrow [(4 \times m_1) \times g] - T = (4 \times m_1) \times a.$ (1)

Were you stuck here? One extra step that you will have to learn now is that for the cylinder,

STEP 2: $\sum \tau =$ net torque $= I \times \alpha$ where I is the moment of inertia and α is the angular acceleration. It is the tension acting off the axis of rotation that produces the torque, but in this step it is enough to realize there is a net torque.

$\Rightarrow T \times R = I \times \alpha \Rightarrow T = [(I \times \alpha) / R].$ (2)

STEP 3:   Now we need to connect a the linear acceleration with α the angular acceleration.

Note that $a = (R \times \alpha).$ $\Rightarrow \alpha = (a / R).$ Do not forget this step!

Go back to equation (2):

The moment of inertia about the axis of rotation $I = [(m_1 \times R^2) / 2].$

Equation (2) now becomes:

$\Rightarrow T = \{[(m_1 \times R^2 / 2)] \times [(a / R)]\} / R = [(m_1 \times a) / 2].$

$\Rightarrow T = [(m_1 \times a) / 2].$ (3)

Add equations (1) and (3): $[(4 \times m_1) \times g] - T + T = [(4 \times m_1) \times a] + [(m_1 \times a) / 2].$

$\Rightarrow 4g = 4a.$

$\Rightarrow a = 4g / 4.5.$

The ratio $4 / 4.5$ is the same as $8 / 9.$

[OR substitute for T and solve for a.

$\Rightarrow$ Equation (1) becomes (substitute for tension):

$[(4 \times m_1 \times g)] - [(m_1 \times a) / 2] = [4 \times m_1 \times a].$

$[4 \times m_1 \times g] = [(4 \times m_1 \times a)] + [(m_1 \times a) / 2].$

Cancel mass m_1 throughout. $\Rightarrow 4 \times g = a \times (4 + 0.5).$

$\Rightarrow a = 4 \times g / 4.5.$

The ratio $4 / 4.5$ is the same as $8 / 9.$]

Answer A!

73.   STEP 1: $\sum \tau =$ net torque $= T \times R = I_{\text{CYLINDER}} \times \alpha.$

STEP 2:   Substitute $\alpha = a / R.$

$\Rightarrow T \times R = I \times (a / R).$

Substitute for $I = [(m_1 \times R^2) / 2].$ Note that the mass of the cylinder is $m_1.$

$\Rightarrow T \times R = [(m_1 \times R^2) / 2] \times (a / R).$

We hope you have not forgotten that the linear acceleration $a = R\alpha$. If you forget that you will be stuck for a while without knowing what to do!

$\Rightarrow$ Net torque $= T \times R = (m_1 \times R \times a) / 2$.

      Note asked: Solve for $T = (m_1 \times a) / 2$.

Answer D!

74.   Angular acceleration, $\alpha = a / R$. Do not forget this relationship. Many students forget it. That is why.
     Answer D!

76.   If an object moves in a straight line with constant acceleration we can always use the equations of kinematics! We hope you have not forgotten that either.

     STEP 1:   Find acceleration using equations of kinematics for constant acceleration.

               Since it is released from rest, we have to use: $\Delta y = (v_{0y} \times t) + (\frac{1}{2} \times a_y \times t^2)$.

               Review the first few chapters. The object is falling below the starting point.

               $\Rightarrow \Delta y$ is negative $= -h$.

               It is released from rest. $\Rightarrow v_{0y}$ is zero.

               Δy is the displacement of the mass $\Rightarrow \Delta y = -2$ m, time $t = 0.8$ s.

               $\Rightarrow -2 = 0 + \frac{1}{2} a \times 0.8^2$.

$\Rightarrow a = -(2 \times 2) / 0.8^2 = -6.25$ m/s^2.

Acceleration is negative for the 20 kg mass because net force is down for the 20 kg mass.

Answer A!

77. $a = R \times \alpha$, where a = linear acceleration, R is the radius and α is the angular acceleration in rad/s^2. The acceleration is 6.25 m/s^2 (by problem **76**)

     $\Rightarrow \alpha = a / R = 6.25 / 0.1$ m $= 62.5$ rad/s^2. That is, the angular speed changes by 62.5 rad/s in one second!

     Note that one rad of apple pie ≈ 59 degrees!

     Answer C!

79. I = moment of inertia of the cylinder $= (m \times R^2) / 2$ about an axis passing through the axis of the cylinder. Unfortunately, the mass of the cylinder is not given. Does this mean that we cannot do this problem? No! Follow these steps very closely!

     There is a net torque (Are you comfortable with this word, torque?) on the cylindrical drum.

     STEP 1:   Net torque $= (I \times \alpha)$.

               We get net torque on an object by the way in which the forces act off the axis of rotation. So the real question is: Can you come up with what is on the left side, net torque?

               The tensions are different for an accelerating cylindrical disc. These are forces. Because they act off the axis of rotation, it will produce a torque.

               $\Rightarrow [(T_2 \times R)]_{\text{BIGGER CLOCKWISE TORQUE}} - [(T_1 \times R)]_{\text{SMALLER COUNTERCLOCKWISE TORQUE}}$

                     $= [(I \times \alpha)]_{\text{NET TORQUE}}$.   **(1)**

STEP 2: $\alpha = (a / R) = 6.25 / 0.1 = 62.5$ rad/s^2.

This step is not easy to come by if you are not used to it. If you miss this step, you will be in trouble. We have used $a = 6.25$ m/s^2 from problem **76**.

STEP 3: For the 4-kg mass, by Newton's Law:

$[T_1]$~BIGGER FORCE~ $- [m_1 \times g]$~SMALLER FORCE~ $= [m_1 \times a]$~NET FORCE~.

$\Rightarrow T_1 - (4 \times 10) = 4 \times 6.25$.

$\Rightarrow T_1 = 65$ N.

STEP 4: For the 20-kg mass:

$[m_2 \times g]$~BIGGER FORCE~ $- [T_2]$~SMALLER FORCE~ $= [m_2 \times a]$~NET FORCE~.

$\Rightarrow (20 \times 10) - T_2 = 20 \times 6.25$.

$\Rightarrow T_2 = 75$ N.

Now we are ready to substitute the quantities that we know back into equation (**1**):

$(75 \times 0.1$ m$) - (65 \times 0.1) = I \times 62.5$ rad/s^2.

$\Rightarrow I = 0.016$ kg $\times$ m^2.

Answer D!

80. We know that $\omega_f = \omega_0 + (\alpha \times t)$.

$\Rightarrow \omega_f = 0 + (\alpha \times t) = (\alpha \times t)$ if the object (cylinder) starts from rest.

Note that ω_0 is the initial angular speed.

We know α = angular acceleration = 62.5 rad/s^2 from problem **77**.

$\Rightarrow \omega_f = (62.5$ rad/s$^2) \times (0.8$ s$) = 50$ rad/s.

Do you know how to convert rad/s to revolutions/s?

Note that 50 rad/s = (50 rad/s) $\times$ [(1 rev)/ (2π radians)] = 7.96 rev/s.

Answer B!

82. The rotational counterpart for force is torque.

Answer B!

83. The rotational counterpart for momentum is angular momentum.

Answer C!

85. Just as net force $= m \times a = \Delta p / \Delta t$, we have net torque $= I \times \alpha = \Delta L / \Delta t$.

If angular momentum L is a constant, net torque is zero.

Answer B!

86. Yes!

Answer A!

88. Normal force acts upward. Tension acts downward at the rim. Weight of the disc acts down.

Answer A!

89.   Since normal force and weight act at the axis of rotation, those forces will not contribute to torque. Tension acts at a distance R from the center. This contributes to torque. *Net torque* $= I \times \alpha = T \times R$.
Answer B!

91.   You need to get used to the idea of calculating acceleration when the object is rotating and coming down the inclined plane.
Let us go with energy conservation. Assume no frictional loss.
If there is no energy input we have: *Total energy at the start = Total energy at the end of the journey.*
$\Rightarrow KE_A + PE_A = KE_B + PE_B$.
A new thing you have to get used to is the rotational kinetic energy.
$\Rightarrow KE_A + PE_A = KE_{B, \text{TRANSLATIONAL}} + KE_{B, \text{ROTATIONAL}} + PE_B$.
$\Rightarrow 0 + (m \times g \times h) = (\frac{1}{2} \times m \times v^2)_{\text{TRANSLATIONAL}} + (\frac{1}{2} \times I \times \omega^2)_{\text{ROTATIONAL}} + 0$.
Note again that KE has rotational and translational "components".
$\Rightarrow m \times g \times h = [\frac{1}{2} \times m \times v^2]_{\text{TRANSLATIONAL}} + [\frac{1}{2} \times I \times \omega^2]_{\text{ROTATIONAL}}$.
For a cylinder, $I_{\text{CYLINDER}} = m \times R^2 / 2$. Also $\omega = v / R$.
$\Rightarrow m \times g \times h = [\frac{1}{2} \times m \times v^2]_{\text{TRANSLATIONAL}} + \{\frac{1}{2} \times [(1/2) \times m \times R^2] \times [(v^2 / R^2)]\}_{\text{ROTATIONAL}}$.
$\Rightarrow [m \times g \times h] = [\frac{1}{2} \times m \times v^2]_{\text{TRANSLATIONAL}} + [(1/4) \times m \times v^2]$. Now cancel mass m.
$\Rightarrow [g \times h] = [v^2 / 2] + [v^2 / 4]$.
$\Rightarrow v^2 \times [(\frac{1}{2}) + (1/4)] = (g \times h)$.
$\Rightarrow v^2 \times [0.5 + 0.25] = (g \times h)$.
$\Rightarrow v^2 = (g \times h) / 0.75$.   **(1)**

    We are not asked for v. What we need is a. Aargh! We are not done yet!
    From equations of kinematics we have $v^2 = v_0^2 + (2 \times a \times l)$.
    $\Rightarrow v^2 = 0 + (2 \times a \times l)$ if the object starts from rest.
    $\Rightarrow v^2 = (2 \times a \times l)$.   **(2)**
    We try to relate this relation to what we got before for v^2 in equation **(1)**.
$\Rightarrow v^2 = (2 \times a \times l) = (g \times h) / 0.75$.
$\Rightarrow (2 \times a \times l) = [(g \times h) / 0.75]$.
$\Rightarrow a = [(g \times h)] / [2 \times 0.75 \times l]$.
$\Rightarrow a = [(g \times h)] / [1.5 \times l]$.
$\Rightarrow a = (g / 1.5) \times (h / l)$.
$h / l = \sin \theta$. $\Rightarrow a = (g / 1.5) \times \sin \theta$.
$\Rightarrow a = (0.67 \ g \times \sin \theta)$. We have used $1 / 1.5$. Note that $0.67 = 2 / 3$.
$\Rightarrow a = (2/3) \times 10 \times \sin \theta$.
Answer D!

92.   Find the final speed at the base. We know the initial speed. Use equations of kinematics.
TO FIND SPEED:
STEP 1: If there is no energy input or frictional loss: Total Energy at A = Total Energy at B.
$\Rightarrow KE_A + PE_A = KE_B + PE_B$.
Do not forget that kinetic energy has two components, translational and rotational.

$\Rightarrow KE_A + PE_A = KE_{B,\,\text{TRANSLATIONAL}} + KE_{B,\,\text{ROTATIONAL}} + PE_B.$

$KE_A = \frac{1}{2} \times m \times v^2 = 0$ if $v_A = 0.$

$PE_A = m \times g \times h = m \times g \times l \times \sin\theta$. Note that $h = l \times \sin\theta.$

$KE_B = KE_T + KE_R$ where KE_T is the kinetic energy due to translation and KE_R is the kinetic energy due to rotation.

$PE_B = m \times g \times h_B = 0$ as height of B from the zero of PE is zero.

$\Rightarrow KE_{B,\,\text{TRANSLATIONAL}} + KE_{B,\,\text{ROTATIONAL}} = m \times g \times l \times \sin\theta$ **(1)**

$KE_{B,\,\text{TRANSLATIONAL}} = \frac{1}{2} \times m \times v^2.$

$KE_{B,\,\text{ROTATIONAL}} = \frac{1}{2} \times I \times \omega^2 = [(I / 2) \times (v / R)^2] = [\frac{1}{2} \times I \times (v^2 / R^2)]$. We used $\omega = v / R.$

$\Rightarrow KE_{B,\,\text{TRANSLATIONAL}} + KE_{B,\,\text{ROTATIONAL}} = [\frac{1}{2} \times m \times v^2] + [\frac{1}{2} \times I \times (v^2 / R^2)].$

$\Rightarrow m \times g \times l \times \sin\theta = (v^2 / 2) \times [m + (I / R^2)].$

Note that we used $\omega = v/R.$ Do not forget that $\omega = v/R.$

Multiply both sides by 2. $\Rightarrow v^2 \times [m + (I / R^2)] = 2 \times m \times g \times l \times \sin\theta.$

For a ring, $I = m \times R^2.$

$\Rightarrow v^2 \times \{m + [(m \times R^2) / R^2]\} = 2 \times m \times g \times l \times \sin\theta.$

This leaves us with:

$v^2 = v_B^2 = g \times l \times \sin\theta$ **(2)**

Note that v_B is the speed at point B.

STEP 2: Now use $v_f^2 = v_0^2 + (2 \times a \times \Delta x).$ **(3)**

If $v_0 = 0$, $v_f^2 = 2 \times a \times \Delta x = 2 \times a \times l$. The slant height of the inclined plane is $l.$

Set the right sides of **(2)** and **(3)** equal.

$\Rightarrow g \times l \times \sin\theta = 2 \times a \times l.$

$\Rightarrow a = g \times \sin\theta / 2.$

Answer C!

94.   Let angular deceleration be $\alpha.$

Use $\omega_f = \omega_0 + (\alpha \times t).$

It is coming to rest due to frictional torque. Once it comes to rest, ω_f = final angular speed = 0.

$\Rightarrow 0 = \omega_0 + (\alpha \times t).$

Note that α has the opposite sign with respect to the angular velocity. It is negative here because of deceleration.

$\Rightarrow 0 = \omega_0 + [(-\alpha) \times t].$

$\Rightarrow \alpha = \omega_0 / t.$ Substitute into net torque = $I \times \alpha.$

Net torque = $I \times \alpha = [(m \times R^2) / 2] \times (\omega_0 / t) = (m \times R^2 \times \omega_0) / (2 \times t).$

Answer B!

95.   Net torque = $I \times \alpha.$

The real question is if you know how to relate α to the number of rotations.

Of course you know. Net torque = $I \times \alpha.$ We know that $\omega_f^2 = \omega_0^2 + (2 \times \alpha \times \Delta\theta).$

If it comes to rest, $\omega_f = 0.$ α is negative if it is decelerating. $\Delta\theta = (2 \times \pi \text{ radians}) \times$ number of rotations

$= 2 \times \pi \times n$ radians.

$\Rightarrow 0 = \omega_0^2 + [2 \times (-\alpha) \times (2 \times \pi \times n)]$.

$\Rightarrow \alpha = \omega_0^2 / (4 \times \pi \times n)$.

The inertia for the flywheel $I = [(m \times R^2) / 2]$.

Net torque $= I \times \alpha = [(m \times R^2) / 2] \times [\omega_0^2 / (4 \times \pi \times n)] = (m \times R^2 \times \omega_0^2) / (8 \times \pi \times n)$.

Answer C!

97. We are asked to find the speed at B. Let us use energy conservation to find speed at point B.

If there is no (negligible) energy loss due to friction, then: Total Energy at A = Total Energy at B.

$\Rightarrow KE_A + PE_A = KE_B + PE_B$.

Do not forget that kinetic energy has two components, translational and rotational.

$\Rightarrow KE_A + PE_A = KE_{B, \text{TRANSLATIONAL}} + KE_{B, \text{ROTATIONAL}} + PE_B$.

Note that at the start, the object was starting from rest, and so kinetic energy at the start is zero.

$KE_A + PE_A = KE_B + PE_A$ (1)

$KE_A = \frac{1}{2} \times m \times v_A^2 = 0$ as $v_A = 0$ if it started from rest at point A.

Note that $\sin \theta = $ [opposite side to θ / hypotenuse] $= [h / l]$.

$\Rightarrow h = l \sin \theta$.

$PE_A = m \times g \times h_A = m \times g \times l \times \sin\theta$.

$KE_B = KE_{B, \text{TRANSLATIONAL}} + KE_{B, \text{ROTATIONAL}}$.

$KE_T = $ Translational kinetic energy $= \frac{1}{2} \times m \times v_B^2$.

$KE_R = $ Rotational kinetic energy $= \frac{1}{2} \times I \times \omega^2$.

Moment of inertia $I = [(m \times R^2) / 2]$.

Also, you need to know that $\omega = [v_B / R]$.

$KE_R = $ Rotational kinetic energy $= \frac{1}{2} \times I \times \omega^2 = \frac{1}{2} \times [(m \times R^2) / 2] \times [v_B / R]^2 = (m \times v_B^2) / 4$.

Now substitute back into equation (1).

$\Rightarrow 0 + (m \times g \times l \times \sin \theta) = (\frac{1}{2} \times m \times v_B^2) + [(1/4) \times m \times v_B^2] + 0$.

Note that the potential energy at B is zero. Cancel mass m through out.

$[g \times l \times \sin \theta] = v_B^2 \times [(\frac{1}{2}) + (1/4)]$.

$\Rightarrow [g \times l \times \sin \theta] = [v_B^2 \times (0.75)]$.

$\Rightarrow v_B^2 = [(g \times l \times \sin \theta)] / 0.75$.

$\Rightarrow v_B^2 = [(g \times l \times \sin \theta)] / (3 / 4)$. We used $0.75 = 3 / 4$.

$\Rightarrow v_B^2 = (4g \, l \, \sin \theta) / 3$. The speed at B is the square root of this number.

Answer A!

98. $KE_B = PE_A$ if it starts from rest. Also, $PE_A = mgh = mgl \sin \theta$.

Remember that $TE_A = TE_B$.

$\Rightarrow KE_A + PE_A = KE_B + PE_B$. This gives us: $KE_B = mgh = mgl \sin \theta$.

[Note that at point B, $PE_B = 0$. Also do not forget that kinetic energy has two components: translational and rotational. $KE_B = KE_{B, \text{TRANSLATIONAL}} + KE_{B, \text{ROTATIONAL}}$.]

Answer B!

100. The component of the weight acting down towards the base of the plane parallel to the inclined surface
$= m \times g \times sin\, \theta$.
Answer B!

101. Angular momentum, $L = I \times \omega = [(m \times R^2)/2] \times \omega$..
Do not forget that $\omega = [v_B / R]$. All we have to do is find the final speed.
To find v_B, use energy conservation.
If there is no energy loss or energy input, then:
$TE_A = TE_B$. $\Rightarrow KE_A + PE_A = KE_B + PE_B$.
$\Rightarrow PE_A = KE_B$.
Again we are reminding you that kinetic energy at B has two components. One is translational and the other
is rotational. $PE_A = m \times g \times h_A$.
$\Rightarrow m \times g \times h_A = KE_T + KE_R$.
Note that the potential energy is zero at the finishing point.
$\Rightarrow m \times g \times l \times sin\, \theta = [\frac{1}{2} \times m \times v_B^2] + [\frac{1}{2} \times I \times \omega^2]$.
$\Rightarrow m \times g \times l \times sin\, \theta = [\frac{1}{2} \times m \times v_B^2] + \{\frac{1}{2} \times [(m \times R^2)/2] \times (v_B / R)^2\}$.
Cancel mass m through out.
$\Rightarrow g \times l \times sin\, \theta = (v_B^2 / 2) + (v_B^2 / 4)$.
$\Rightarrow g \times l \times sin\, \theta = v_B^2 (0.5 + 0.25) = v_B^2 \times 0.75$.
$\Rightarrow (g \times l \times sin\, \theta) / 0.75 = v_B^2$.
$\Rightarrow v_B = [(g \times l \times sin\, \theta) / 0.75]^{1/2}$.　　**(1)**
Angular momentum, $L = I \times \omega = [(m \times R^2)/2] \times \omega$, where $\omega = v_B / R$.
$\Rightarrow \omega = (v_B / R) = (1 / R) \times v_B = 1 / R \times [(g \times l \times sin\, \theta) / 0.75]^{1/2}$.
$\Rightarrow L = I \times \omega = [(m \times R^2)/2] \times \{1 / R \times [(g \times l \times sin\, \theta) / 0.75]^{1/2}\}$.
$\Rightarrow L = I \times \omega = [(m \times R)/2] \times [(g \times l \times sin\, \theta) / 0.75]^{1/2}$. Now use $3/4 = 0.75$.
$\Rightarrow L = I \times \omega = [(m \times R)/2] \times [(4 \times g \times l \times sin\, \theta) / 3]^{1/2}$.
Answer C!

103. Psych it out: $I_{CYLINDER} > I_{SPHERE}$. This means that the acceleration of the cylinder is less than the
acceleration of the sphere. That is, the ratio of their accelerations is less than 1.
Only one choice is less than 1.
For the cylinder:
$TE_A = TE_B$.
$\Rightarrow KE_A + PE_A = KE_B + PE_B$. $\Rightarrow PE_A = KE_B$.
$m \times g \times l \times sin\, \theta = [\frac{1}{2} \times m \times v_B^2] + [\frac{1}{2} \times I \times \omega^2]$.
$\Rightarrow m \times g \times l \times sin\, \theta = [\frac{1}{2} \times m \times v_B^2] + \{\frac{1}{2} \times [(m \times R^2)/2] \times (v_B / R)^2\}$.
$\Rightarrow m \times g \times l \times sin\, \theta = [0.75 \times m \times v_B^2]$.
$\Rightarrow v_B^2 = (g \times l \times sin\, \theta) / 0.75$.
From equations of kinematics, $v_B^2 = v_A^2 + [2 \times a \times \Delta x]$.
a is the acceleration of the cylinder $= a_c$. At point A, speed $v_A = 0$ m/s.
$\Rightarrow v_B^2 = [(g \times l \times sin\, \theta) / 0.75] = 0 + (2 \times a_c \times l)$. Cancel l and rearrange.

$\Rightarrow a_c$ = acceleration of the cylinder = $(g \times sin\ \theta) / (2 \times 0.75) = (g \times sin\ \theta) / 1.5$ **(1)**.

For the sphere:

$TE_A = TE_B$.

$\Rightarrow KE_A + PE_A = KE_B + PE_B$.

$\Rightarrow PE_A = KE_B$.

$\Rightarrow m \times g \times l \times sin\ \theta = [\frac{1}{2} \times m \times v_B^2] + [\frac{1}{2} \times I \times \omega^2]$.

Substitute $I = \frac{1}{2} \times [(2/5) \times m \times R^2$ for a sphere.

$\Rightarrow m \times g \times l \times sin\ \theta = [\frac{1}{2} \times m \times v_B^2] + \{\frac{1}{2} \times [(2/5) \times m \times R^2] \times (v_B / R)^2$.

$\Rightarrow m \times g \times l \times sin\ \theta = [\frac{1}{2} \times m \times v_B^2] + [(1/5) \times m \times v_B^2]$.

$\Rightarrow m \times g \times l \times sin\ \theta = 0.7 \times m \times v_B^2$. We used: $\frac{1}{2} + (1/5) = 0.7$.

Cancel mass m.

$\Rightarrow v_B^2 = [(g \times l \times sin\ \theta) / 0.7] = v_A^2 + [2 \times a_s \times (\Delta x)]$.

Speed at point A, v_A is zero.

$\Rightarrow v_B^2 = [(g \times l \times sin\ \theta) / 0.7] = 0 + [2 \times a_s \times l]$.

$\Rightarrow v_B^2 = [(g \times l \times sin\ \theta) / (2 \times 0.7)] = a_s \times l$. Cancel l.

$\Rightarrow a_{SPHERE} = (g \times sin\ \theta) / 1.4$ **(2)**

From the results of **(1)** and **(2)** we get:

$\Rightarrow a_{CYLINDER} / a_{SPHERE} = [(g \times sin\ \theta) / 1.5] / [(g \times sin\ \theta) / 1.4]$

$= 1.4 / 1.5 = 0.93$.

Answer B!

104.   The ratio of their velocities at point B will be proportional to the square root of the accelerations.

$v_f^2 = v_A^2 + (2 \times a \times l)$.

$\Rightarrow v_B^2 = 2 \times a \times l. \Rightarrow v_B =$ square root of $(2 \times a \times l) = (2 \times a \times l)^{1/2}$.

For the cylinder:

$TE_A = TE_B$.

$\Rightarrow KE_A + PE_A = KE_B + PE_B$.

$\Rightarrow PE_A = KE_B$. Note that speed at point A is zero.

$\Rightarrow m \times g \times l \times sin\ \theta = (\frac{1}{2} \times m \times v_B^2) + (\frac{1}{2} I \times \omega^2)$.

Substitute $I = (m \times R^2 / 2)$, and $\omega = (v_B / R)$.

$\Rightarrow m \times g \times l \times sin\ \theta = (\frac{1}{2} \times m \times v_B^2) + [\frac{1}{2} \times (m \times R^2 / 2) \times (v_B / R)^2]$.

$\Rightarrow m \times g \times l \times sin\ \theta = 0.75 \times m \times v_B^2$ or $v_B = [(g \times l \times sin\ \theta) / 0.75]^{1/2}$.   **(1)**.

For the sphere:

$TE_A = TE_B$.

$\Rightarrow KE_A + PE_A = KE_B + PE_B$.

$\Rightarrow PE_A = KE_B$.

$\Rightarrow m \times g \times l \times sin\ \theta = (\frac{1}{2} \times m \times v_B^2) + (\frac{1}{2} I \times \omega^2)$.

Substitute $I = (2 \times m \times R^2 / 5)$, and $\omega = (v_B / R)$.

$\Rightarrow m \times g \times l \times sin\ \theta = (\frac{1}{2} \times m \times v_B^2) + [\frac{1}{2} \times (2 \times m \times R^2 / 5) \times (v_B / R)^2]$.

$\Rightarrow m \times g \times l \times sin\ \theta = 0.7 \times m \times v_B^2 \Rightarrow v_B = [(g \times l \times sin\ \theta) / 0.7]^{1/2}$.   **(2)**.

From **(1)** and **(2)** we get the ratio of their velocities to be:

$v_{CYLINDER} / v_{SPHERE} = (0.7 / 0.75)^{1/2}$.

$\Rightarrow v_{CYLINDER} / v_{SPHERE} = 0.97$.

Answer B!

106.   *PE* for both the cylinder and the sphere $= m \times g \times l \times sin\ \theta$.

        $\Rightarrow$ Their ratio = 1. We like this question!

        Answer C!

107.   Angular momentum, $L = I \times \omega = I \times (v / R)$.   **(1)**

        For the cylinder:

        $TE_A = TE_B$.

        $\Rightarrow KE_A + PE_A = KE_B + PE_B$.

        $\Rightarrow PE_A = KE_B$.

        $\Rightarrow m \times g \times l \times sin\ \theta = \frac{1}{2} \times m \times v_B^2 + [\frac{1}{2} \times (m \times R^2 / 2) \times (v_B / R)^2]$.

        $\Rightarrow v_B = [(g \times l \times sin\ \theta) / 0.75]^{1/2}$.

        $\Rightarrow$ Equation **(1)** becomes $L_1 = \{(m \times R^2 / 2) \times [(g \times l \times sin\ \theta) / 0.75]^{1/2} \} / R$.

        For the sphere:

        $TE_A = TE_B \Rightarrow PE_A = KE_B$.

        $\Rightarrow m \times g \times l \times sin\ \theta = (\frac{1}{2} \times m \times v_B^2) + [\frac{1}{2} \times (2/5 \times m \times R^2) \times (v_B / R)^2]$.

        $\Rightarrow v_B = [(g \times l \times sin\ \theta) / 0.7]^{1/2}$.

        Angular momentum, $L = I \times \omega = I \times (v / R)$

        Equation **(1)** becomes $L_2 = \{(2/5 \times m \times R^2) \times [(g \times l \times sin\ \theta) / 0.7]^{1/2}\} / R$.

        $\Rightarrow L_1 / L_2 = (0.5 / 0.4) \times (0.7 / 0.75)^{1/2} = 1.2$.

        Answer B!

109.   If there is no friction, objects do not roll but slide down or slip.

        Answer B!

110.   For a point on the rim, velocity changes (in direction). That is, the momentum changes. For the center of mass, velocity is zero.

        Answer A!

112.   Since the angular momentum is a constant, net torque $= \Delta L / \Delta t = 0$.

        Answer B!

113.   Let T_1 be the tension between m_2 and m_1; T_2 the tension between m_1 and m_3.

For the mass m_2 accelerating up: $T_1 - (m_2 \times g \times sin\ \theta) = m_2 \times a.$ **(1)**

For the mass m_3 accelerating down: $(m_3 \times g) - T_2 = m_3 \times a.$ **(2)**

Add **(1)** and **(2)** and we get: $T_1 - T_2 - (m_2 \times g \times sin\ \theta) + (m_3 \times g) = (m_2 + m_3) \times a.$

Rearranging we get: $\{g \times [m_3 - (m_2 \times sin\ \theta)]\} - (T_2 - T_1) = (m_2 + m_3) \times a.$ **(3)**

For the pulley, net torque $= I \times \alpha = (T_2 - T_1) \times R.$

Substitute: $I = (m_1 \times R^2) / 2,$ and $\alpha = a / R.$

$\Rightarrow I \times (a / R) = (T_2 - T_1) \times R.$

$\Rightarrow [(m_1 \times R^2) / 2] \times [(a / R)] = (T_2 - T_1) \times R.$

$\Rightarrow (T_2 - T_1) = (m_1 \times a) / 2.$

$\Rightarrow \{g \times [m_3 - (m_2 \times sin\ \theta)]\} - [(m_1 \times a) / 2] = (m_2 + m_3) \times a.$

$\Rightarrow \{g \times [m_3 - (m_2 \times sin\ \theta)]\} = [m_2 + m_3 + (m_1 / 2)] \times a.$

$m_3 = 4 \times m_1, m_1 = m_1, m_2 = 2 \times m_1.$

$g \times [(4 \times m_1) - (2 \times m_1 \times sin\ \theta)] = [(2 \times m_1) + (4 \times m_1) + (m_1 / 2)] \times a.$

We are told that: $\theta = 30.$

$\{g \times [(4 \times m_1) - (2 \times m_1 \times sin\ 30)] = [(2 \times m_1) + (4 \times m_1) + (m_1 / 2)] \times a.$

$3 \times g = 6.5 \times a.$

$\Rightarrow a = (3 \times g) / 6.5 = 30 / 6.5 = 4.6\ m/s^2.$

Answer A!

115. $6 \times T = m \times g.$

$\Rightarrow T = [(m \times g) / 6].$

You can make the pulling force small if there are more pulleys.

Answer B!

1.   By definition of simple harmonic motion, the acceleration is proportional to the displacement from the fixed point O and is always directed *towards* the equilibrium point O. Net force felt by the mass m is due to the spring.

$F_{\text{NET FORCE ON THE MASS}} = m_{\text{MASS}} \, a_{\text{ACCELERATION}} = -k_{\text{SPRING CONSTANT}} \, x_{\text{DISPLACEMENT FROM EQUILIBRIUM}}$.

Hence, acceleration $a = -(k/m) \times x$ where $k/m \, (= \omega^2)$ is a constant for a particular spring with a mass attached to it.

Answer A!

2.   The amplitude $A_{\text{AMPLITUDE}}$ is the maximum displacement of the particle on either side of the equilibrium point O and hence $A_{\text{AMPLITUDE}} = OA_{\text{maximum displacement}} = OB_{\text{maximum displacement}} = $ maximum displacement from the equilibrium point.

Answer C!

4.   The velocity is maximum at equilibrium point O. Note that the acceleration at the equilibrium point is zero.

Answer D!

5.   Period $= 2\pi \times (m/k)^{1/2} = 2\pi/\omega$.

Note that $k/m \, (= \omega^2)$ is a constant for a given spring and mass. That is, the period is the same for small amplitudes or large.

Answer C!

7.   Point A is the turning point where the magnitude of the acceleration is maximum but speed is zero.

Answer A!

8.   Point B is the turning point where the magnitude of the acceleration is maximum but speed is zero.

Answer A!

10.   Kinetic energy of the mass in joules $= (1/2)m_{\text{MASS}} \, v_{\text{SPEED}}^2$.

This is zero if speed is zero (at the turning points).

Answer A!

11.   Kinetic energy of the mass in joules $= (1/2)m_{\text{MASS}} \, v_{\text{SPEED}}^2$. This is non-zero if speed is non-zero.

Answer B!

13.   Period $= 2\pi \times (m/k)^{1/2} = 2\pi/\omega$.

Note that $k/m \, (= \omega^2)$ is a constant for a spring. The period does depend upon the mass.

A greater mass gives a greater (larger) period for a fixed spring constant.

Answer A!

14. Period $= 2\pi \times (m / k)^{1/2} = 2\pi / \omega$. Note that $k / m \, (= \omega^2)$ is a constant for a spring.
Period depends upon the spring constant. A greater spring constant gives a smaller (shorter) period.
Answer A!

16. Potential energy PE of the system in joules $= (1/2) \, k_{\text{SPRING CONSTANT}} \, x_{\text{DISPLACEMENT FROM EQUILIBRIUM}}^{\,2}$.
At point A, displacement is maximum. That makes PE non‑zero.
In this particular case, $PE = (1/2) \, k_{\text{SPRING CONSTANT}} \, A^2$.
Answer B!

17. Potential energy PE of the system in joules $= (1/2) \, k_{\text{SPRING CONSTANT}} \, x_{\text{DISPLACEMENT FROM EQUILIBRIUM}}^{\,2}$.
At point B, displacement is maximum. That makes PE non‑zero.
In this particular case, $PE = (1/2) \, k_{\text{SPRING CONSTANT}} \, A^2$.
Answer B!

19. Potential energy (PE) of the spring in joules $= (1/2) \, k_{\text{SPRING CONSTANT}} \, x_{\text{DISPLACEMENT FROM EQUILIBRIUM}}^{\,2}$.
If $x_{\text{DISPLACEMENT FROM EQUILIBRIUM}}$ is non‑zero, PE is non‑zero.
Answer B!

20. $[TE]_{\text{TOTAL ENERGY}}$ in joules $=$
$\quad\quad [KE = (1/2) m_{\text{MASS}} \, v_{\text{SPEED}}^{\,2}]_{\text{Kinetic Energy}} \, +$
$\quad\quad\quad\quad [PE = (1/2) \, k_{\text{SPRING CONSTANT}} \, x_{\text{DISPLACEMENT from EQUILIBRIUM}}^{\,2}.]_{\text{Potential Energy}}\cdot$
At equilibrium point displacement from equilibrium is zero. This makes the contribution to PE zero. Or TE is all kinetic at point O.
Answer A!

22. At a point like P that is between points A and B, we have both kinetic energy and potential energy because there is a non‑zero displacement from the equilibrium point and a non‑zero speed.
Answer A!

23. $[TE]_{\text{TOTAL ENERGY}}$ in joules $=$
$\quad\quad [KE = (1/2) m_{\text{MASS}} \, v_{\text{SPEED}}^{\,2}]_{\text{Kinetic Energy}} \, +$
$\quad\quad\quad\quad [PE = (1/2) \, k_{\text{SPRING CONSTANT}} \, x_{\text{DISPLACEMENT FROM EQUILIBRIUM}}^{\,2}.]_{\text{Potential Energy}}\cdot$
Both KE and PE are positive or zero, and their sum can never take a negative value.
Answer B!

25. Frequency does not depend upon the amplitude.
Frequency $f = 1 / T_{\text{PERIOD}}\cdot$
Period $= 2\pi \times (m / k)^{1/2} = 2\pi / \omega$. Note that $k / m \, (= \omega^2)$ is a constant for a spring.
Answer B!

26. Manufacturers make springs of different spring constant.
What is constant about a spring is its spring constant k.
Answer A!

28. Hertz = $1/s$.
Answer A!

29. Because of the square term involving speed and displacement, we do get two similar points. That is, it does not matter if the displacement is $+x$ or $-x$.
Answer A!

31. The individual contributions may vary but the sum will always be a constant if there is no friction. Note that:
$[TE]_{\text{TOTAL ENERGY}}$ in joules =
$$[KE = (1/2)m_{\text{MASS}} \, v_{\text{SPEED}}^2]_{\text{Kinetic Energy}} +$$
$$[PE = (1/2) \, k_{\text{SPRING CONSTANT}} \, x_{\text{DISPLACEMENT FROM EQUILIBRIUM}}^2 \cdot]_{\text{Potential Energy}}.$$
Answer A!

32. The spring force will always oppose the stretching. Hence the force with which spring acts on a mass = $k \times x$ opposite to the direction of stretching.
Answer A!

34. The amplitude is A.
Answer C!

35. The kinetic energy = $[\frac{1}{2} \times m \times v^2] = TE - PE =$
$[\frac{1}{2} \times k \times A^2]_{\text{TOTAL ENERGY}} - [\frac{1}{2} \times k \times x^2]_{\text{POTENTIAL ENERGY}}$
$= (\frac{1}{2} \times k \times A^2) - (\frac{1}{2} \times k \times x^2)$.
Note: For simple harmonic motion, the velocity is: $A \times \omega \times [1 - (x^2/A^2)]^{1/2}$,
where $\omega = 2\pi/T = (k/m)^{1/2}$.
Hence: $v_{\text{SPEED}} =$
$$A_{\text{AMPLITUDE}} \times \{(k_{\text{SPRING CONSTANT}}/m_{\text{MASS}})^{1/2} \times [1 - (x_{\text{DISPLACEMENT}}^2/A_{\text{AMPLITUDE}}^2)]^{1/2}\}.$$
Answer A!

37. The kinetic energy = $(\frac{1}{2} \times m \times v^2) = TE - PE =$
$(\frac{1}{2} \times k \times A^2)_{\text{TOTAL ENERGY}} - (\frac{1}{2} \times k \times x^2)_{\text{POTENTIAL ENERGY}}$
$= (\frac{1}{2} \times k \times A^2) - (\frac{1}{2} \times k \times x^2)$.
Answer C!

38. The potential energy $= \frac{1}{2} \times k \times x^2$.

    Note: The potential energy = total energy – kinetic energy
    $$= (\frac{1}{2} \times k \times A^2)_{\text{TOTAL ENERGY}} - \{\frac{1}{2} \times k \times A^2 \times [1 - (x^2/A^2)]\}_{\text{KINETIC ENERGY}} = \frac{1}{2} \times k \times x^2.$$
    Answer D!

40. Total energy is conserved.
    Answer A!

41. The force is not a constant. It depends upon the displacement from the equilibrium point.
    Answer B!

43. $T = 2\pi \times$ [square root of (m/k)] $= 2\pi \times$ square root of $\{m/[(m \times g)/l]\} = $
    $2\pi \times$ square root of (l/g). Note that $[(m \times g)/l] = k$.
    [Again, note that $mg = kl$.]
    Answer C!

44. Amplitude $= A$.
    Answer D!

46. Since kinetic energy is always positive it changes only from a maximum to zero and hence its frequency is
    twice that of the oscillation, i.e., $2 \times \{(1/2\pi) \times$ [square root of (g/l)]$\} = (1/\pi) \times$ [square root of (g/l)].
    Answer B!

47. Since the elastic potential energy ($= \frac{1}{2} \times k \times x^2$) is always positive, it changes only from a maximum to zero.
    Hence the frequency of variation of potential energy is twice that of the oscillation, i.e.,
    $2 \times \{(1/2\pi) \times$ [square root of (g/l)]$\} = (1/\pi) \times$ [square root of (g/l)].
    Answer C!

49. Force on each spring serving to stretch it is the same $= m \times g$.
    Answer A!

50. Total stretching is the sum of the stretchings of s_1 and s_2, i.e., $[(m \times g)/k_1]_{\text{FIRST}} + [(m \times g)/k_2]_{\text{SECOND}}$.
    Answer D!

52. The stretching of each spring is the same $= l$.
    Answer B!

53. Force on $s_1 = k_1 \times l$ and that on $s_2 = k_2 \times l$.
    For symmetrical mass m to lie horizontally, $k_1 l = k_2 l$, so $k_1 = k_2$ and the problem is one of two identical
    springs.
    Answer C!

55.   Note that $k_{\text{EFFECTIVE}} = k_1 + k_2$.
Stretching = Force$/k_{\text{EFFECTIVE}} = (m \times g)/(k_1 + k_2)$.
Answer D!

56.   Force $= m \times g$.
Effective spring constant of two springs in series is given by :
$1/k_{\text{EFFECTIVE}} = 1/k_1 + 1/k_2$.
$\Rightarrow k_{\text{EFFECTIVE}} = (k_1 \times k_2)/(k_1 + k_2) = k^2/(k + k) = k/2$.
Stretching = Force$/$effective spring constant $=$
$(m \times g)/k_{\text{EFFECTIVE}} = (2 \times m \times g)/k$.
Answer A!

58.   Effective spring constant $k = 16 \times k$.
Frequency $= 1/(2\pi) \times$ [square root of $(16k/m)$] $= 1/(2\pi) \times$ [square root of $(16 \times k/m)$]
$= (2/\pi) \times$ [square root of (k/m)].
Answer A!

59.   When m is displaced to right, k_1 is stretched and k_2 is compressed.
Restoring force on m, is $(-k_1 \times x) + (-k_2 \times x) = -(k_1 + k_2) \times x$.
The force is $(k_1 + k_2) \times x$ to the left.
Answer B!

61.   Answer D!

62.   For each half of the folded spring, the effective spring constant is $2 \times k$ and the folded system consists of two springs each of spring constant $2 \times k$ in parallel. Hence the effective spring constant $= 4 \times k$.
Answer D!

64.   Each mass will oscillate about the other.
The reduced mass μ given by $1/\mu = (1/m_1) + (1/m_2)$.
$\mu = (m_1 \times m_2)/(m_1 + m_2)$.
Period $T = 2\pi \times$ [square root of (μ/k)] $= 2\pi \times$ {square root of $[(m_1 \times m_2)/(m_1 + m_2) \times k]$}.
Answer B!

65.   Reduced mass $\mu = (m_H \times m_{Cl})/(m_H + m_{Cl}) = (35/36) \times m$.
$f = n = (1/2\pi) \times$ [square root of (k/μ)].
Square both sides and solve for $k = 4\pi^2 \times \mu \times n^2$.
Substitute the values of HCl into $\mu = (m_1 \times m_2)/(m_1 + m_2)$.
$\Rightarrow f = n = 4\pi^2 \times [(35/36) \times m] \times n^2 =$
$(35/9) \times \pi^2 \times n^2 \times m$.
Answer A!

CHAPTER 10, Solution Page **5**

67. If k is the spring constant, $2 \times m \times g = k \times l$ and $k = (2 \times m \times g)/l$.
Total mass moved $= m + (2 \times m) = 3 \times m$.
Period of oscillation $= 2\pi \times \{$square root of $(3 \times m)/k\} = 2\pi \times \{$square root of $[(3 \times l)/(2 \times g)]\}$.
Answer D!

68. No! The acceleration is always toward O, so acceleration and velocity vectors are in toward O, so in opposite directions. This will reduce the speed.
Answer B!